The Complete Book of Oscilloscopes

2nd Edition

The Complete Book of Oscilloscopes
2nd Edition

Stan Prentiss

TAB BOOKS

Blue Ridge Summit, PA

SECOND EDITION
FIRST PRINTING

© 1992 by **TAB Books**.
TAB Books is a division of McGraw-Hill, Inc.

Printed in the United States of America. All rights reserved. The publisher takes no
responsibility for the use of any of the materials or methods described in this book,
nor for the products thereof.

Library of Congress Cataloging-in-Publication Data

Prentiss, Stan.
    The complete book of oscilloscopes / by Stan Prentiss. — 2nd ed.
        p.   cm.
    Includes index.
        ISBN 0-8306-3909-8 (hard)   ISBN 0-8306-3908-X (paper)
        1. Cathode ray oscilloscope.   I. Title.
    TK7878.7.P697   1992
    621.381′5483—dc20   91-32801
                                                                              CIP
                                                                              r91

TAB Books offers software for sale. For information and a catalog, please contact
TAB Software Department, Blue Ridge Summit, PA 17294-0850.

Acquisitions Editor: Roland S. Phelps
Book Editor: Andrew Yoder
Production: Katherine G. Brown
Book Design: Jaclyn J. Boone
Cover Design: Cindy Staub, Hanover, PA   EL2

Contents

Acknowledgments   *ix*

Introduction   *xi*

❖   **1   Oscilloscopes of all persuasions**   **1**

The tactical approach   *3*
   Applications   *4*
Instrument capabilities   *6*
   DSOs   *8*
   Logic analyzers   *11*
   Spectrum analyzers   *13*
   Analyzer types   *16*
   Waveform analyzers   *19*
   Oscilloscope cameras vs plotter/printers   *23*

❖   **2   Waveforms and their analysis**   **27**

As waveforms develop   *28*
The square wave   *30*
Waveshaping   *32*
Pulses help too   *36*
Coupling circuits   *37*
Filters   *40*
Video   *43*
The NTSC signal   *44*
Test signals   *47*
VITS and VIRS   *50*

❖   **3   The extraordinary oscilloscope**   **54**

How they work   *55*
   The basic scope   *57*
   The scope's important time base   *65*

Smart scopes w/printouts   *70*
   The FX-850 Epson   *75*
Applications   *79*
   HM 604 examples   *81*
   Analog scopes can have digital readouts   *88*

❖ **4 The bits and bytes of digital storage**   **96**
Keep it simple   *97*
Historically speaking   *98*
Technicalities   *100*
Tektronix' contribution   *104*
   Routine analog   *106*
   Exciting storage   *108*
Hewlett-Packard   *113*
   Measurements   *115*
   Glitches and single shots   *115*
   Initial examples   *116*
   Wrap up   *120*
Troubleshooting   *124*

❖ **5 Our wonderful world of spectrum analyzers**   **126**
Analyzer topics and selections   *127*
   Modestly-priced equipments   *129*
Analyzer applications   *137*
   Signal-to-noise (terrestrial and satellite)   *137*
   Baseband analysis   *145*
   Noise   *145*
Looking at frequency response   *150*
   Multichannel sound   *154*
The high end   *158*
   Modulation   *159*
   The 492PGM analyzer   *162*
   Action in high places   *164*
   New Tektronix offering   *170*

❖ **6 Logic analysis in state and timing logic**   **172**
Why analyzers?   *173*
Easier is often better   *175*
   Hewlett-Packard   *176*
   Practical aspects   *179*
   Triggering in state analysis   *179*
Innovators   *181*
   Fluke/Philips   *181*
   Characteristics   *186*
   Scope or analyzer   *188*

State or timing   *188*
Data acquisition and analysis   *191*
Disassembly   *193*

❖ 7   **Time-domain reflectometry**   **194**

Basics and more   *195*
Complex loads   *198*
Test signals   *200*
Large reactances   *201*
Small reactances   *204*
Padders   *204*
Circuit-board connectors   *205*
Long-line TDRs   *208*
TDR setup and applications   *209*
Millirho   *215*
Optical reflectometers   *217*
Fiber-optic specifics   *218*
Connections   *220*
Installations   *221*
ODTR applications   *221*
Definitions   *226*
Valuable tips   *226*

❖ 8   **The laboratory of tomorrow**   **229**

No hangar flying   *230*
GPIB and RS-232   *232*
Objectives   *234*
An IBM interface   *235*
Hi-res audio I/O board for Mac II   *236*
Software   *239*
LabVIEW   *239*
LabWindows   *241*
Futurebus +   *243*

❖ 9   **Signature analysis**   **245**

Logical numbers   *245*
Decimal radix   *246*
Binary   *247*
Binary coded decimal   *248*
Octal   *248*
Hexadecimal   *249*
ASCII   *250*
EBCDIC   *250*
Data transfer   *254*
Sync   *255*

Modulation/demodulation   *255*
A digital circuit tester   *256*
Signature characters and operation   *257*
Accuracy   *260*
Noise   *262*
Manual troubleshooting   *263*

❖ **10   Vectorscopes and vectors**   ***265***
How vectors originate   *269*
Lissajous patterns   *274*
Calculating phase shifts   *280*
Illustrating phase shifts   *282*
Vectors by sidelock generation   *285*
Applications   *290*
Summary   *293*

Index   ***295***

Acknowledgments

WE GRATEFULLY EXPRESS OUR DEEP APPRECIATION TO ALL THOSE who have aided in this second edition of *The Complete Book of Oscilloscopes*. Without their more than considerable aid, we could not have accomplished the objective of technical information that is so important for a well-rounded and accurate description of these many instruments and systems discussed throughout the text.

David Grabove and Eric Williams of Hewlett-Packard; Tim Dehne and Roxanne Wied of National Instruments; Curtis Smith, Robert Oblack, Lynn Hurd, Larry Haga, Roger Lucas, and Len Garrett of Tektronix; Mark Mays, and Tak Tsang and Debby Coyne of John Fluke Mfg.; Mike Silverstein of Hameg; Michael McCown of BK-Precision; Jeffrey Lawson of Polaroid; John Heitman of Advantest; Tom Dideum of IFR; M. Dale Sherrill of Comm/Scope; George Birch of Nicolet; Mark Zirngast of LeCroy; and Dennis Raulin of GTE Spacenet.

If any names have been omitted, it's pure oversight. However, all are remembered with more than simple gratitude from your author.

Introduction

THIS SECOND EDITION OF *THE COMPLETE BOOK OF OSCILLOSCOPES* is almost a completely new book in every respect. Instruments, applications, techniques, descriptions, and troubleshooting information pertaining to this equipment and the hardware/software that they serve has been carefully researched and chronicled for maximum use in this expanding, diverse and highly instrumented world of electronics. If you possess the first edition of *The Complete Book of Oscilloscopes*, the combined volumes should cover a considerable number of both product offerings and utility over much of the past decade and serve as worthwhile reference material for even longer. From the first book, only one chapter has been reprinted in its entirety, and eight of the ten chapters in this one are completely new and fully state-of-the-art.

Of major interest is the extensive material that is devoted to digital storage oscilloscopes (DSOs); spectrum analyzers (from dc to 21 GHz); logic analyzers (with the new Fluke/Philips state and timing modes exhibited via a single probe on one screen), and others by Hewlett-Packard; a very special chapter on metal cable and fiberoptics time-domain reflectometry; particular information on storage and A/D/A converters; progressive and new applications of ordinary oscilloscopes; and especially an explanation of GPIB, RS-232, and Futurebus instrument connectors, digital printers; and a detailed description of what to expect in the laboratory of tomorrow.

Chapter 8 even explains how you can configure your own test equipment, complete with knobs and connections to troubleshoot or design special electronics wanted or existing that require extraordinary methods or attention. Software and hardware are explained in considerable detail, based on information furnished directly by a generous manufacturer. We have even included additional knowledge on signature analyzers in the expectation that extraordinary complexity in large-scale integrated circuits (LSI) will eventually force some manufacturers to adopt node logic readouts, rather than system or subsystem on-screen displays, which can and do leave much to be desired.

Overall, both text and illustrations should appropriately justify the title of this and its preceding sister volume, and furnish a wealth of useful information for all in the world of digital and analog electronics.

❖ 1
Oscilloscopes
of all persuasions

OSCILLOSCOPES WERE ONCE A RELATIVELY EASY TOPIC TO explain. They were either *recurrent* or *triggered sweeps*; 25 megahertz (MHz) was high frequency; time bases from seconds to microseconds with 5% accuracies were phenomenal, and transistorized equipment was in the ascendancy over current-hogging, high-voltage vacuum tubes. Now, the only tube remaining in the usual oscilloscope is the display device and that, too, might see its demise when liquid-crystal displays (or other flat-panel readouts) become legible, higher-frequency realities. Whenever these changes occur, the pictures on the wall won't be confined to simply television and computer readouts. All sorts of video and measurement test parameters will bring resolution and precision that we can only fantasize about today—all in multi-color images for instant recognition and convenient comparison. In tube technology, among the more expensive laboratory instruments of today, many of these objectives have already been reached, but only corporations with extensive R & D can normally afford the price. Us lesser folk will have to suffer with 60- to 150-MHz instruments that might not have all the Christmas decorations, but do have 2-3% accuracies, nanosecond time bases, between two and four-channel inputs, very fast trace risetimes, adjustable markers (called *cursors*), programmable measurements, on-screen readouts, counter-timers, voltmeters, auto-tests, and relative or exceptional portabilities (depending on scope complexities).

Although bigger is not necessarily better, the larger oscilloscopes usually do pack considerably more functions within their metal skins than those designed for a small suitcase or briefcase, but at a heftier price tag. Thus, the differential is between $1000 to 2000 and $3000 to 4000. Above $4000, you should have some extremely important measurements to access or accommodate that pay fairly fast dividends to recover such considerable outlays. Certified, calibrated rentals might be one answer for the usual oscilloscope, but not for some other equipment that we'll cover later. The reason is that complex instruments require familiar operators; equipment becomes more comfortable and accurate over time. Spectrum and waveform analyses are two disciplines where ownership and skilled, trained personnel are decidedly required. Therefore, you must precisely understand an instrument's mission and purchase accordingly. At Goddard Space Flight Center (NASA), we often had the various sales representatives, and even factory engineers, trot out their wares for a "full dog and pony show" before any genuine procurement document was ever written. Then, if one piece of equipment was clearly superior (especially one where more-than-normal funding was required), we wrote a sole-source request that would stand the test of competitive challenge.

Some corporate and government institutions are imbued with the "lowest bidder" or the open-market approach, but this won't work advantageously where an unusual and specific need arises to investigate and verify some special parameter, artifact, or condition. It is also important to remember that design engineers depend heavily on top-quality test gear to prove their software or hardware calculations and construction. Many printed circuit (PC) boards or semiconductors don't operate according to design because some incidental glitch has been overlooked or neglected. One of the most important functions in original design is the debugging thereof. With today's complexities, thousands of gates, flip-flops, logic and linear combinations, metallizations, and multiple input/outputs (I/Os) just don't respond perfectly the first time. If anyone is under the impression that they do, trot down to Silicon Valley and interview a dozen or so harassed engineers with big-college degrees and discover for yourself. Difficult problems are always begging for solutions. That's why expert debuggers are priceless! Sometimes it pays to be shop-smart and a little lucky. Conceptually, the EE is great, but common sense and logic/linear knowledge often rescue many a valiant project for everyone, including the EE.

I can't help remembering one experience when a young engineer, fresh out of college, hung three current meters on the base, emitter, and collector of a transistor to discover its collective operation with signal inputs. Either basic calculations or a dual-trace oscilloscope with probes would have proved all in much less time than a flock of 5% meters hung on unshielded electrodes. It made the young man appear both juvenile and rather foolish. One suspects he never heard of biasing and saturation; most institutions of higher learning concentrate on digital, rather than ordinary linear signal processing. Thus, many EEs grope around the insidious paths of electron and hole flow on their solitary snipe hunts, with or without the bag.

The tactical approach

Training is the next general subject that's sorely needed in the measurement industry, and only high-quality reps and factory sales "engineers" seem qualified. All the money in the world spent on classy-chassis equipment won't produce usable results, unless an operator can both read and interpret most numbers and displays. This is why no-cost seminars by industry giants, such as Tektronix and Hewlett-Packard, are totally invaluable. Thereafter, individual attention might be needed if instrument complexities or special applications reveal further help. We'll bear all this in recallable memory as various approaches are discussed. Spectrum analysis is the most neglected category and therefore will shortly become a prime topic, accompanied by interesting and instructive results. It's no use designing systems that won't operate in intended environments, whatever the objective. Impedances and drive voltages/currents must readily mate for practical products, and system couplings can easily result in a 3-dB (a 50-percent) power loss—a factor that becomes devastating under small signal conditions (especially in 5-volt ICs). Regardless of all careful calculations, computer-aided designs (CADs), experienced engineers, and medium- and high-frequency systems (especially) should have their operating spectrums evaluated both during design and under all operating conditions, for both logic (if used) purity and/or signal processing.

All of the foregoing sounds rather easy and somewhat paternalistic, but you can't imagine the disappointments in an ever-so-carefully laid out series of nodes and branches or digital logic chips when an entire project refuses to work. Then, even steel doors might not deter a raving project director. But good, solid basic training does, and coupled with experience, ordinary hypertension is often controlled without an expensive jug of pills. The boss might have to take only an occasional blood-thinning aspirin to avoid an imminent coronary.

The best, most serious advice that we can offer consists of factory-training routines, application notes, careful evaluation of all instrument parameters, comparison with competing units, and sales-engineer aid, when and where required. With new instrument introductions almost monthly, the old days of annual announcements have evaporated, and rivalry is strong. Select the best, most cost-effective measuring device, check all of its functions, then begin the process of learning all that you can from whatever source is handy. We'll even contribute as much as we can where appropriate, and even more in ordinary measurements where the need seems greatest.

Applications

A generalized discussion of applications becomes difficult when no evident goal, breakdown, or system analysis is available. However, several principles worth noting and understanding can be applied to almost any electronic problem.

- If there is an output, work backwards toward the source.
- No output indicates open, short, or defective power supplies. The latter should be investigated first, for without sufficient voltage and current, electronic systems will not operate. However, such voltage should be measured under load.
- Check the input. A sloppy drive voltage or logic contamination will often cause initial receiving circuits to further compound the problem because they're not designed to operate under such conditions.
- Discover or devise a signal block diagram (if that's possible) and attempt to trace from the input toward the output.
- Where multiple subsystems are connected, try disconnecting one at a time to see if a portion of the system responds.
- In display systems, the cathodes of the picture tube will often indicate a telltale reason for no picture. The lack of high voltage (second anode voltage) can produce the same effect.
- Unfortunately, if the cathode(s) lacks video drive voltage, the tube will also show no picture.
- Digital bit streams are simply 1s and 0s. If these aren't

present, a hot-to-touch IC suggests either a short or high-current drain from some other device.

- Bit-stream errors, of course, distort the output and result in garbled transmissions. But don't confuse hexadecimal, binary, and other allied logic, or your application will be garbled too.
- Heat will force even stabilized systems into errors or break-downs. When ambient temperatures are above 90°F, shut it down or add extra air conditioning (if available). Even venting closed equipment will help.
- Above all, use nonloading instrumentation to search for trouble. Remember, a 10-kΩ restive probe shunted across a 10-kΩ circuit delivers only a 5-kΩ resistance to common (ground). When using oscilloscopes, 10X low-capacitance probes with 10-MΩ shunt impedance are always preferred.
- High-frequency equipment, such as spectrum analyzers, are normally set for 50 or 75 Ω, matching the circuit Zs they're measuring. However, many newer oscilloscopes and especially older spectrum analyzers are dc volts-sensitive and won't tolerate large, critical overloads. Once again, know your measuring instrument characteristics. Repair and calibration charges are often excessive, and instruments that have internally stored calibration standards, although somewhat more costly, can actually save money in a few years. This is especially true of spectrum analyzers where shop time becomes as valuable as a chunk of Fort Knox or the Philadelphia Mint.

For starters, the previous list should move troubleshooting thought processes forward a few bytes or volts and serve as a guide to seemingly lesser problems that are often more difficult to discover than the majors. Many subtleties do not commonly reoccur and therefore can't be tagged as repetitious. However, current, voltage, and logic measurements of various descriptions usually will find the culprit lurking in some elusive shadow. Here, perseverance and patience are your right-hand assistants as you probe here and there with digital and analog instruments. If readouts require accuracies to three places or more, by all means use equipment with digital readouts—especially for digital volt-meters with analog bar graphs. For instance, 5-V logic will certainly require considerably higher accuracy than those that require 20 V. We prefer 4.5-digit meters with 100% overranging and automatic scaling; these meters save much burnout and repair. Furthermore, for maximum accuracy, oscilloscopes and voltmeters (especially ac types), should have two times the frequency response over the parameters they're testing (just for good measure, of course). The same does not necessarily hold true for electronic counters, but 12-place readouts do improve accuracies in addition to dual 50-Ω and 1-MΩ input impedances.

Instrument capabilities

Whether test gear meets your present and forseeable future needs has a great deal to do with its usefulness. A small or inexperienced business will attempt to "get by" with outmoded, uncalibrated instruments that are derived from various sources only because they're cheap. This way is the easiest that we know of to lose customers and market share. If your products or services fail advertised objectives or are even marginal, you're a prime candidate for bankruptcy. It's always better to borrow or underwrite better-quality instruments (even though the process is painful), rather than fail for want of a few hundred or several thousands more.

For instance, you're looking at two or four channels of analog or logic that has clock frequencies or bandwidths of 50 MHz. Is that all the passband that is required of your oscilloscope? Hardly . . . sync and trace triggering might be sufficient, but many oscilloscopes offer "linear" traces of only 60% of their ranges and "tail off" afterwards. How do you do this? Simply hook up a constant voltage generator (often used for calibrations) to an input scope channel, adjust generator to a convenient seven-graticule division and crank up the frequency until the trace barely fills four divisions (Fig. 1-1). This is the 3-dB downpoint where 70% of the display remains. Notice that even this display is not linear. So, by ratio and proportion, you'd need 100 MHz to measure 50 MHz linearly:

$$60{:}36 = 100{:}x \quad \text{or } 60x = 3\,600, \text{ and } x = 60 \text{ (MHz)}$$

Now suppose, also, that your dc parameters range from 5 to 10 V/div. You're aware that ordinary oscilloscopes have 8 vertical divisions and 10 horizontal. So, if you had to look at almost 150 volts with a direct probe, you couldn't do it. With a 10x low-capacity probe 5 $\times$ 8 $\times$ 10 equals 400 volts, you could. For a true, general-service oscilloscope, we recommend 20 V/div.,

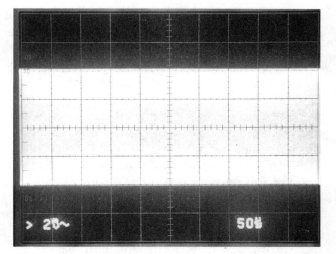

Fig. 1-1 *The scope's − 3-dB passband is evident as RF envelope decreases from 7 to 4 divisions with inputs from a constant-voltage generator.*

which calculates out to 1 600 volts (10 ×), ac plus the dc component. Unfortunately, as a result of the ascendancy of digital logic, most scopes today offer only 0.2 to 5 volts per division, so you usually must use a 10 × or 100 × probe, like it or not.

This section is simply a quick peek into oscilloscope parameters that will be expanded considerably when full explanations of specific instruments are in order. Some cathode-ray tube equipment is tricky, so you should have maximum information available for best selection. In addition, several varieties of pure oscilloscopes have very specific uses, rather than one instrument for everything. These will be carefully explained also and a number of illustrative examples are included.

Remember that oscilloscopes without graticule lighting are difficult to ''read'' in dimly lighted work rooms. Tired eyes aren't conducive to extra accuracy, therefore oscilloscopes without CRT illumination are to be avoided under all but the most unusual circumstances. If you are planning fast risetime and falltime measurements, vertical amplifier bandwidths and time bases in the nanoseconds are highly desirable. Once again, accuracy of such tests depends on the scope's passband being at least double that of the parameters to be investigated. For example, a 35-microsecond risetime would require scopes with vertical amplifier bandwidths of:

$$Tr = 350/\text{bandwidth or } 35 \times 10^{-6} = 350/\text{bandwidth}$$
$$\text{and bandwidth} = 350/35 \times 10^{-6} = 10 \text{ MHz}$$

If the *Tr* was 35 nanoseconds, then your scope would require a bandwidth of 10 GHz. Comprende? Beware of the rolloff factor and be sure your time base reads accurately in at least high nanoseconds for the first and low nanoseconds for the second. Bandwidths and time bases, as you can see, are inseparable for accurate waveform evaluations, although they are not necessarily equivalent.

Recalling the accuracy factor; for a 5% error, an ordinary oscilloscope requires a 3:1 ratio of passband to that of the equipment it's measuring. For a 1% accurate reading, you'll need a 7:1 ratio. In other words, the scope must exceed the frequency/passband of the measured unit by 3 times in the first instance, and 7 times in the second. So for 1% accuracy of a 20-MHz signal, a 150-MHz oscilloscope would just about cover the action. True, ratios have been improved in the very newest scopes and we'll talk more about specific measurements as various types of waveshapes are covered. However, these instruments are not the dime-store variety, and several thousand dollars is required to acquire a premium instrument sufficient for better-than-average measurements. Another factor arises in that most of the better oscilloscopes have analog vertical and horizontal accuracies of 2%, and the lesser instruments no better than 3% (even when brand new). And as they age, 3% can easily become 5%—especially if annual calibration is carelessly neglected.

Fortunately, many of the newer oscilloscopes have both frequency counters and digital voltmeters built in, with the added advantage of on-screen readouts. Further, time-base cursor markers have now been added for additional accuracy and special timing measurements. The once laborious task of counting vertical and horizontal graticule divisions and calculating amplitudes and times is now reduced to simply viewing the digitally-supplied on-screen information and either committing it to memory or recording the result by a dot-matrix or X-Y recorder printer. In short, the better instruments today have little relationship to even those of a few years ago. Considering increased accuracies, overall performance, and 3- and 5-year guarantees (which translate into reliability), the cost/production product isn't significantly different, and might even be less.

DSOs

So far we've really only discussed analog scopes in a very general way, neglecting one of the more important logic advances in the measurements industry: that of *digital-storage oscilloscopes*

(DSOs). Like all else, this equipment has good points and bad, so you must know how to use them for maximum effectiveness. In addition, many of today's DSOs also have standard analog functions with vastly better passbands than their digital portions. In most DSOs, divide the megabit sampling rate by a factor of 10 for the actual digitized bandwidth. As an example, a 100-MHz analog readout versus a 100-Mbit rate would actually display 10 MHz digitized. Although this gross rule of thumb might not necessarily apply in top-of-the-line instruments, ADA converters have to become considerably faster and less expensive for the ordinary lower frequency DSOs. Therefore, sometime during the 1990s, high-frequency DSOs will cost a considerable penny.

As you might have guessed, oscilloscope cathode ray tubes for analog and digital scopes are either similar or identical because both functions are normally available in any recent package. However, where the analog scope's time base simply sweeps and triggers incoming information from the vertical amplifiers, DSOs first analog-to-digitally convert analog waveforms into logic bit patterns and send them to sequential memories for storage. For display, these bit patterns are read out serially and are routed to a D/A converter, which develops an analog display on the CRT's vertical deflection plates and screen comparable to the scope's analog input.

From this description it's easy to understand that higher sampling rates result in more detailed images and faster reproductions. Some DSOs use equivalent time sampling for higher frequency signals that exceed their normal sampling frequencies. Sampling only a portion of the input signal during one sweep, such a process can reconstruct a fast waveform by adding samples together, forming a composite readout. This process is not possible, however, unless the signal is totally recurrent. So as each signal period arrives, a trigger pulse generates a ramp whose amplitude is compared with that of the output staircase-generating circuit. When both signals are equal, the generator output increments to a higher value and finally forms the completed waveform.

One of the prime advantages of such scopes is the ability to examine events, such as trigger points, before and after the waveforms of interest. Controls can also be set to retrieve selected information in storage before the triggering event. But the greatest digital scope advantage is its ability to capture and hold single-shot events, which its pure analog relative can't possibly do.

However, at least one analog scope recently introduced to the market at a comparatively reasonable price will actually do more than a digital scope at intermediate speeds. This scope can also print the results on an inexpensive dot-matrix printer. The only exception to this claim, however, is the single-shot, nonrecurring voltage, which requires absolute storage for accurate reproduction. But analog scopes still lead the race in storage bandwidth, as a result of both sampling techniques and displays. The Nyquist sampling rate requires two or more samples per waveform cycle—anything less will produce *aliasing* (a lower frequency distorted version of the original information). Therefore, digitized signals need more than two samples per cycle at the highest frequency. Thus, digitizing must be very rapid to reproduce accurate signals. You might also encounter occasional horizontal jitter and/or sample density, both as a result of violating the Nyquist rule. Expanded waveforms can also have problems if digital design is at all faulty.

As you already know, costs are considerable in designing and producing a faster-than-average digitizing oscilloscope. Consequently, such instruments are expensive when fast megasamples/second are required.

According to practical mathematicians, the useful bandwidth (USB) of any storage oscilloscope amounts to the following as originally conceived:

$$\text{USB} = \frac{\text{digitizing rate (in MHz)}}{25 \text{ (for dot displays) or } 10 \text{ (for vector displays)}}$$

That's about all the introduction to DSOs you should need. We'll save the remainder for the chapter on the real, live equipment in action, followed by an intensive explanation of spectrum analysis in digital readouts. Even though one tries, it's probably impossible to overemphasize the importance of these two oscilloscope categories. Neither is actually new, but recent engineering developments and specific applications have made all varieties of analyzers highly important to both design and debugging of the many electronic devices today. So, be prepared for a couple of larger than normal chapters as we continue. Some remarkable information is available for everyone. We'll try to keep it both readable and straightforward even though some aspects will be intensively technical.

Logic analyzers

Somewhat kith and kin to the DSOs, the logic analyzer can be virtually whatever you want and what you're willing to pay. Eight, 16, 32, 64 channels, etc., are available to almost any reasonable number you'd care to name, although the price is proportional to the complexity. Wherever digital circuits are, the logic analyzer has certainly become the general-purpose instrument for bit-stream investigations, design aid, data acquisition and display separation, trigger or clock analysis, connection multiple ''slices'' of information, word lengths in various chunks or bytes of logic, communications, and equipment system functions (Fig. 1-2).

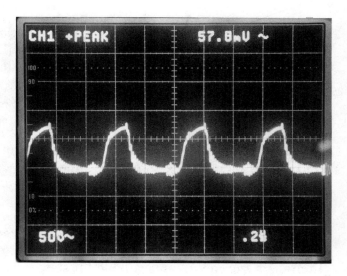

Fig. 1-2 *A very suspect 2-MHz bitstream, which illustrates overshoot, ringing, poor symmetry, and a low amplitude of only 57.8 millivolts.*

Initially, incoming information translates into binary (1s and 0s) and is then memorized, regardless of the number of inputs. Readouts can then be compared, displayed at various trigger rates, sampled, and separated; ICs and discrete transistor actions are examined; risetimes, glitch captures, and dropped bits identified—some or all of which might be handled synchronously (with clock) or asynchronously (without clock). The output then becomes some sort of channel-word format for CRT display. Such analyzers might also have both external triggers and external clocks, depending on the need. Triggers are especially important to start and stop data flowing out of memories upon command. In this way, both transients and glitches can be identified—even if they occur sequentially in either slow or rapid timing. Both real time and sample modes are always available under operator control.

Newer logic analyzers have both state and timing functions. In "state," analyzers compress incoming data into numbers or characters (such as binary, hexadecimal, ASCII, or octal), and they can also investigate microprocessors—in short, a software analyzer. In "timing," multiple logic channels are captured and displayed as timing diagrams. Therefore, each function is highly important in working with the more complex equipment. In software, you develop the program; in timing, you operate the hardware.

Once again, instrument clock speeds play a vital part. For effective electronic troubleshooting, analyzer clock speed/resolution has 10 times that of the *device under test* (DUT). Accumulating memory is another equally important parameter that can be influenced by data storage on every clock pulse or just during 1s and 0s transitions. As an example, analyzers that operate between 400 and 500 MHz can capture 2 to 2.5-nanosecond glitches. However, reducing speeds to 100 MHz might or might not offer 5 nsec reliability. Transitional timing, on the other hand, stores data during transition times only, compresses this data by eliminating redundancies, thereby increasing information in available memory.

Selectable post- and pretriggering modes also affect memory, because advanced pretriggering can allow a total memory scan for further pretrigger examination. Glitch-capture latches and glitch memories are often available for storage and reuse, which operate automatically, but are sometimes obscured when glitches appear on or as sample points. The cure is simply to summon a faster sampling rate to increase the analyzer's resolution.

Even the best logic analyzer can be upset by undesirable input impedances that load; and also by timing skew, which results from using low resistance; high-capacitance probes, which should approximate 10 MΩ and 8 pF. Thereafter, equipment-skew specifications should appear as no more than 2 to 3 nanoseconds.

A disassembly-display mode can be made available in the more advanced analyzers to select microprocessor op codes for prefetched cycles and to descramble them for real-data instructions.

We could also talk about test vectors and signature analyzers, but the foregoing should be enough to swallow for the moment. The logic analyzer, however, has highly regarded practical applications for not only debugging hardware but it will happily monitor and debug real-time software in microprocessor equipment. Read-only memory (ROM) systems are especially vulnerable, because instructions are fixed in memory at a specific address. In the state mode, a number of events might be recognized, but triggers must detect independent operations within the several states and trigger on a sequence as each event occurs. Newly acquired equipment problems can also be analyzed with specific data storage by the analyzer and returned to the factory of origin for study and solutions. Halt and write signals are always recognizable and they help find errors within specific data streams, especially when related to a timing diagram. Such procedures not only verify trouble areas, but also speed up system and subsystem repairs immensely. Although the analyzer's state mode samples data and multisignal inputs that involve grouping and display of data between ICs and systems, as well as selective bit-stream triggering, its timing mode reveals critical control functions and any interrupts or glitches that might become apparent as the system or ICs process general or specific information.

Don't expect, however, that any logic analyzer is a cure for all flagging systems and their problems. Complex analysis requires superior equipment, and the cost rises proportionally. Also, movements are afoot to combine (integrate) oscilloscopes with logic analyzers, because waveform and state or timing displays should be seen together for maximum information. Apparently, the quest for digital discoveries never ends, and the next generation of logic analyzers will combine timing and state data with a single set of probes, medium-to-high megahertz operations and more than several kilobits of memory/channel at comparatively reasonable prices.

Spectrum analyzers

A highly important category for many varieties of measurements involving primarily Y amplitudes in decibels ($1/10$ of a Bel) and X parameters in frequency. Ordinary oscilloscopes, of course, read out signals in Y volts and X time bases in seconds through microseconds and nanoseconds. Consequently, spectrum analyzers, such as logic and several other types of analyzers, are specialized instruments unto themselves (Fig. 1-3). Already indications are,

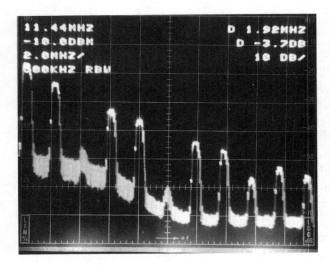

Fig. 1-3 *This is how Fig. 1-2 appears on a spectrum analyzer with a center frequency of 11.44 MHz, a resolution bandwidth of 500 kHz, and a frequency readout of 1.92 MHz.*

however, that digital storage oscilloscopes might offer an option to also accommodate spectral analysis, but probably not yet at the higher frequencies or extended vertical "windows," despite considerable additions to the initial price.

For now, we do have a 500-MHz ancillary addition for oscilloscope readouts that is reasonably attractive, and a 1991 standalone unit that is capable of receiving 1 gigahertz signals at the top end and kilohertz information at the bottom. Its "window" display range can't possibly compare with the super analyzers and several other limitations confine this analyzer's measurements to primarily service uses, rather than special laboratory applications, where ranges and tolerances are considerably tighter as a result of precision requirements. Consequently, for those who want the best, cash or credit outlay can easily amount to between $30 and $100 thousand, depending on the manufacturer and some particular instrument. Conversely, those with very special needs will have to pay the piper, while servicers can rather quickly amortize the several thousand dollar outlay by charging a little extra for appropriate use.

Meanwhile, extolling the virtues of an "inexpensive" spectrum analyzer isn't an especially delightful persimmon to chew, because limitations are glaringly apparent, and accuracies always suspect—either vertically or horizontally. For maximum utility and precision, a microprocessor-driven unit should be blessed with a number of menus and references from which to choose. For instance, if you're measuring carrier-to-noise from some RF signal, you must also account for video/audio band-spreads, filters, resolution bandwidth, dc reference/input, on-screen readouts with cursors, calibration in both dBs and frequency, in addition to some other parameters that could be critical to your needs. This example only shows that a special "log cabin" is still a "fer/fur piece" down the road for all those who desire calibrated results. For an "inside-the-park home run," exact measurements might not necessarily be mandatory, but a long ball over the left field fence could establish a deserved place in the world book of records. Similarly, microwave signals in 10^9 GHz frequencies can be extremely important when "firing" across the U.S. or overhead from 22,300 miles in the Clarke Belt. As time and the 20th century draws to a close, you will find that there's room for virtually any measuring instrument that can produce fairly decent readouts of specific phenomena. The frequency and amplitude spectra of so many signals are simply networks of communications, just like neurons and synapse in the human brain. And for these, a multitude of instruments are needed!

Now, with the advent of considerably improved YIG oscillators and much faster analog-to-digital and digital-to-analog (ADAC) converters, both analyzer performance and costs have entered a considerably better-balanced ratio. In other words, a bigger bang for your buck! Signals that seem to disappear in regular oscilloscope time domain are often easily recognized and measured in the frequency domain—especially distortions, oscillations, rings, and similar effects that wouldn't necessarily appear in analog scope time domains. Additionally, if such signals can be viewed and examined in a dual perspective, such as both time and frequency domains, very little is left that you don't know about that particular voltage. Even then, the individual instrument specifications that apply to any spectrum analyzer are worth considering.

Low operational distortion, range, stability, resolution, accuracy, amplitude displays, practical control layout, accelerating voltage, selective on-screen menus, input impedance(s), perhaps portability, and a strong metal shield and cover are all important inducements. Serviceability might be a final factor, especially those instruments with critical ICs in sockets and various printed circuit boards that might be easily removed and turned over to a designated repair shop for restoration or substitution. Poor solder joints or even one leaky IC can often shut down an instrument.

Analyzer types

Analyzers with the most rapid responses are those with *parallel filters* and individual detectors because, once tuned, they can immediately process and display signals within a discrete band of frequencies. Unfortunately, each filter has a finite bandwidth and it becomes significantly expensive to keep adding filters and detectors to cover any considerable spread of frequencies. For occasional rapid or one-shot phenomena, such instruments are most appropriate—especially for the lower-valued signals as a result of both limited resolution and passbands. For parallel-filter analyzers, the cost/utility factor all depends on applications. Apparently great for audio or other low-value analog signals, the cost and technical difficulties for extended RF viewing among other considerations would be prohibitive.

Fast Fourier Transform (FFT) instrument design is now making significant progress, both in speed and spectral range. If you can bear the expense, the more recent FFT's overcome much of the frequency and resolution difficulties that are associated with the parallel filter types, but are still generally limited to the kHz and low MHz ranges. Mathematically, the Fast Fourier Transform translates time domain information into that of the frequency domain and it can also process low Hz information with considerable accuracy and resolution. Time, frequency, and signal phase are all FFT skills that make such instruments extremely useful. Selling prices, however, often range into five or more digits. Frequency limitations, though, should be less as 12-bit (or better) A/D converters become considerably faster and even more accurate. In the foreseeable future, FFT instruments should move into the higher MHz ranges, while still maintaining a considerable number of lines of resolution, as well as an acceptable dynamic range between 60 and 80 dB. A 600-line analyzer that measures frequencies within 10 kHz, for instance, would have a resolution of 10 000/600, or 16.67 Hz/line. Therefore, calculate your requirements accordingly, then inquire about the price. Newer and better integrated circuits can reduce costs appreciably in the next several years.

Tuned Filter analyzers offer an escape from the multiple filter types but they, too, have specific limitations that aren't always attractive. Often used in microwave measurements, swept-tuned instrument filters do have resolution and sensitivity problems at low frequencies and often might not maintain the required filter shape and bandwidth that is needed over considerable spectra spreads. Such an analyzer operates by slowly sweeping its single filter over frequencies of interest and delivers such signals to a detector where required amplification traces its pattern on the cathode ray tube. Simple in design, tuned filter analyzers consist of a tunable bandpass filter, a detector for vertical image detection, and a horizontal scan generator that synchronizes the tuned frequency to the analyzer's horizontal deflection. Resolution is a function of filter bandwidth (often not constant) and, therefore, is dependent on frequency. However, this equipment is relatively inexpensive and entirely suitable for limited measurements.

Superheterodyne analyzers (Fig. 1-4) are more common types of analyzer instruments that have wide frequency ranges, excellent accuracies, and are available in both bench and portable instruments. They have fixed bandpasses, are completely tunable over the specified range, peak-detect the rms value of a sine wave, and sweep the input spectrum across a fixed passband filter. Said by most to be a narrowband superheterodyne receiver, electrically tuned to frequencies via a sawtooth drive to control a portion of a voltage-tuned LO local oscillator, their characteristics range from kHz to GHz. Operationally, they are much like specially-constructed, high-sensitivity radios that are tuned

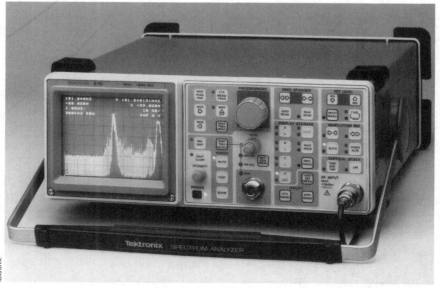

Tektronix.

Fig. 1-4 *A prime example of a 10-kHz to 1.8-GHz microprocessor-controlled superheterodyne analyzer, a model 2710.*

manually or remotely to certain signals and frequencies. Their resolution results from the bandwidth of their intermediate frequency filters that follow RF downconversion from their 50- or 75-Ω inputs.

Swept superhets have extensive "window" amplitude displays, considerable sensitivities, excellent signal resolution, stability, and wide range. Frequency response becomes a difference between an input signal and that of the local oscillator:

$$Rt = F(lo) - F(IF)$$

So, an IF that might be specified at 100 MHz and a LO that tunes 100 to 200 MHz would produce a tuning range $Rt = 100 - 200$, an Rt of 100 MHz.

Many of the latest spectrum analyzers are not only microprocessor controlled, but also feature digital filtering—especially those with selectable IF passbands that results from -3-dB filter bandwidths (the narrower the bandwidth, the better the resolution). Harmonic distortion is another worry resulting from frequencies generated by the local oscillator (LO) and signal mixer, especially third-order harmonics. Where mixer-level third-order problem amplitudes equal the fundamental, this becomes the *third-order intercept (TOI)* and it is very important in calculating any and all mixer distortions that result from TOI. As an example:

$$Distortion\ in\ dBc = 2 \times mixer\ level - 2 \times TOI$$

And should you want to work this out with signal levels into the mixer from, say, -100 to -20 dBm with a given TOI of 10 dBm, you'll find that as signal levels decrease, so does the distortion. Algebraic addition signifies this, and the more negative dBs mean that much less distortion.

Just to illustrate how one factor affects another in these analyzers, instrument noise relates directly to IF filter bandwidths, and many manufacturers specify internal noise as a function of such bandwidths. Therefore, narrow IF filter bandwidths are directly proportional to an analyzer's resolution bandwidth. And resolution bandwidth can be defined as the 3-dB bandwidth of the final IF amplifier. A pair of closely-spaced and equal-amplitude signals are resolved if equivalently separated by the measured analyzer's 3-dB bandwidth. If such does not occur, the two signals will appear as one and your ResBW will be inaccurate.

Today's analyzers tune both frequency and amplitude for

either highly accurate or somewhat relative measurements. Amplitudes are determined in either dB/division or linearly up/down the same vertical scale. Time and frequency measurements might appear horizontally as max. scan, per division, or zero scan. In the older models, a tunable marker indicated fairly exact signal frequencies (as opposed to today's on-screen readouts, cursors, etc.), and recently developed tracking generators supply a signal that exactly tracks tuning of the spectrum analyzer. In *max scan mode*, tracking output is a "start-stop" sweep; in *scanning per division* it's a delta frequency sweep; and in *zero scan*, there's simply a stand-alone continuous signal. Tracking signals develop from mixing a single-oscillator output in the tracking generator and additional oscillator sinewaves from the analyzer itself. Open loop and closed loop are the two tracking generator operating modes.

Special consideration in working with all spectrum analyzers is their *sensitive inputs*. Some newer models have guarded inputs that tolerate dc volts as high as 100 Vdc. Older models require a capacitor input to prevent burning up the first local oscillator and a repair bill that can approach $1,000! Just a tiny high-value capacitor is dearer than a nugget of precious gold!

Waveform analyzers

In some or many respects, waveform analyzers are combinations of oscilloscopes and spectrum analyzers that are capable of displaying analog and sometimes analog/digital signals in both frequency and time-domain spectra. Often quite expensive and rather complex, such instruments are becoming more widely produced as a result of dedicated usage, high accuracy, and being especially designed for the lower frequencies. As design technology continues, waveform analyzers are expected to produce direct readouts in both time and frequency with no cursors or calculations when reproducing ordinary waveforms and their characteristics. As technology improves, waveform analyzers are finding additional uses in automated tests for go/no-go conditions or with extra time-domain processing, pulse analysis, and even multiple-waveshape displays. They're also equipped for single, as well as multiple waveform, processing and analysis, including multiplication, division, addition, subtraction and various correlations, besides calculus operations (such as derivatives and integrals) are possible as well as sinx/x and linear interpolation. In some instances, even an auto-correlation operation is

available to separate random and periodic signals, plus enhanced
storage features, make such an instrument invaluable in factory
QC/QA as well as in the test lab. Be prepared to spend at least five
figures, possibly more, for a good one.

Hewlett-Packard has an HP 3562A Dynamic Signal Analyzer
(Fig. 1-5) with many of the above features at a competitive price.
The instrument is designed for accurate, high-resolution frequency measurements of electronic and mechanical systems via
FFT (Fast-Fourier Transforms), log. and swept sine resolutions,
as well as an internal signal source for random noise and sinewaves. The 3562A is actually a dual-channel spectrum analyzer
with resolution to 26.5 μHz over the 64-μHz to 100-kHz measuring range, with single-channel accuracy of ±0.15 dB within an
80-dB dynamic range. It will handle up to five "auto-sequence"
programs and execute them on command from internal storage
and expand such storage with an outboard disk drive. In its net-
working mode, the instrument offers complete HP-IB programmability. In its digital mode, simultaneous/1 – 2 channel
sampling produces an accuracy of ±0.1 and ±0.5°. Mixed digital and analog signals offer accuracies of ±0.2 and ±2°.

Competing waveform analyzers do equivalent measurements, either manually controlled or programmed, and are especially useful for filter and network investigations, machinery or

Fig. 1-5 *Hewlett-Packard's HP3562A Dynamic Signal Analyzer using fast-
Fourier transforms (FFT) for high-resolution measurements.*

other mechanical vibrations, structural analysis and acoustics. Hewlett-Packard claims that their FFT instruments are typically 10 to 100 times faster than the ordinary swept-frequency front-end analyzers which, of course, are very wide-band, all-purpose, and often cover spectrums from kHz to GHz – something that fast Fourier transforms cannot do.

We could continue this discussion with further specifications and a few interesting innovations, but because waveform analyzers are constantly being developed to accommodate extended applications, much of what is contemporary today could be obsolete tomorrow. Therefore, study your specifications and fit the instrument to the immediate need. Better yet, rent an instrument if experimentation is necessary or the device is only required for a short term. Do not expect an analyzer to perform satisfactorily without a reasonably extensive memory.

Don't, however, sell FFTs short or long until you thoroughly comprehend their advantages and limitations. First, a small memory will certainly restrict spectral resolution. Known as resolution bandwidth on a swept front-end analyzer, FFT spectral resolution depends on the sample period times the number of samples; if memory remains limited, then such measurements can be severely curtailed. Be aware also of the dynamic range of an FFT analyzer: an 8-bit A/D will offer only 48 dB, and a 10-bit A/D only 60 dB, but a 12-bit A/D produces a 72-dB dynamic range. Therefore, when dealing with any instrument that depends on analog-to-digital converters, carefully evaluate the bit-converters before opening the pocketbook.

You must also examine the frequency range. For instance, a conversion rate of 200 megasamples/second will only resolve *maximum* frequencies of 100 MHz; less, if the usual sample rate is decreased. Periodic signals can also result in phase discontinuities under certain circumstances at either end of the display. However, as developments such as voltage and current analysis zoom expanding features, and additional modularity are more readily available, these waveform analyzers become important additions to the growing family of CRT and/or flat-screen apparatuses.

What a boon these might offer to consumer-service people if low-cost versions could be developed to adequately handle both audio and video with special troubleshooting programs! High-definition digitally-encoded television (HDTV) might just force such instruments on the U.S. market by or before 1995. Additionally complex ICs and special signal decoders could, indeed, result in system or subsystem comparison software packages, programmable for the various receivers and essential for any intelligent evaluation or troubleshooting. Could be one outgrowth of fuzzy logic!

We're also noticing more frequent articles in some of the major magazines advocating design-for-testability (DFT) systems that should be part of electronic products—especially circuit and subsystem plug-in PC boards with automatic test equipment (ATE) where applicable and appropriate or simply DFT techniques from startup design. On occasion, however, ATE doesn't do the job as a result of loading, unnecessary test points, and even the recent introduction of surface-mounted RLC and semiconductor devices. Some engineers are already looking at cluster testing and boundary-scan techniques for fault isolations because basic in-circuit examinations require multiple node probes and are often limited to digital logic, rather than a mixture of digital and analog approaches. Unfortunately, we can't offer immediate answers simply because they are not yet available.

Yet, consumer product manufacturers are beginning to offer common-fault listings on computer readouts and others use factory-programmed video, sweep, and allied microprocessor on-screen settings for special or general troubleshooting investigations, even though some TV receivers—especially the large projection sets—are modular (such as Zenith's digital), and either boards or specific large-scale integrated circuits are easily removable. All System 3 Zeniths are modular and divided into RF/video/audio and high/low voltage and are sweep-clustered on two PC boards, with the video amplifiers and drivers for the CRT contained on a third. These receivers, of course, are fully field-serviceable for virtually any breakdown that might occur. Such microprocessor/microcomputer control and system execution will increase rather than diminish, and software troubleshooting techniques are destined to grow with further solid-state sophistication. Inexpensive sets with lesser functions and features could become throwaways well before the next century, just like the monochromes of today. More digital color processing receivers could shortly be on the way—especially because the latest HDTV pure transmission systems are totally digital, with no analog. This topic will be investigated in some detail in a later chapter with TV and satellite waveforms.

In the early 1990s, digital consumer products are just beginning to take shape, with enormous innovations yet to come. One significant test application is the printer-plotter single or dual set of connectors, which is appearing on many of the newer digital-storage oscilloscopes, and even on some analog units.

Oscilloscope cameras vs plotter/printers

With the price of scope film skyrocketing, an interesting comparison can develop between cameras and plotter/printers—especially in the higher frequencies and in circumstances where fast transients are not always apparent, even with relatively reliable storage. At kilohertz repetition rates and fairly clean non-complex traces, either printers or plotters behave satisfactorily and record whatever your scope or analyzer permits. With rapid events—especially those that are nonrepetitive, nothing can replace fast film and an outstanding scope camera. In addition, fuzzy traces (not fuzzy logic) or those with less-than-distinct modulation might be falsely recorded as a different pattern entirely if forced through the usual procedures or poorly designed analog-to-digital converters. Further, insufficient storage in digital scopes or their lack of megabit sampling/second, can distort or entirely ignore waveshapes with faster-than-normal transients or rapid rise and falltimes.

Printer/recorders are not to be condemned; their uses are especially gratifying when signals are relatively straightforward and uncomplicated, but they do suffer considerably in comparison with high-speed types 667 (commercial) or 107 (consumer) flat-pack Polaroid coaterless prints that are listed at 3 000 ISO or DIN numbers (indicating the film's sensitivity to light). In oscilloscope photography, fast film is usually desirable for both maximum resolution and contrast, because slower formulations might not produce sufficient writing-speed characteristics to deliver the best image. We have been working exclusively with Polaroid's highly effective mechanically-triggered and excellent lens DS-34 unit (Fig. 1-6), which has various extension CRT hoods to fit complementary graticules, and the results will be apparent throughout the book. Notice that images are extremely sharp with black/white contrast.

Printers, on the other hand, might automatically operate with large paper loads, print in various type or font settings, produce dedicated-number systems, condense characters or expand, operate in selected formats, do letters and diagrams other than

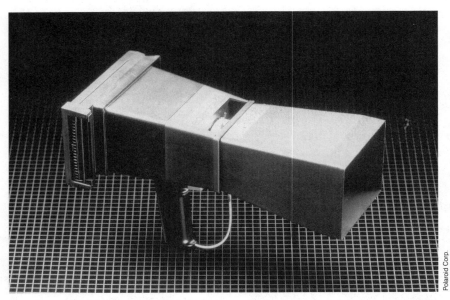

Polaroid Corp.

Fig. 1-6 *Polaroid's DS-34 oscilloscope camera, which has been indispensable for waveform photos that appear throughout the book.*

simply oscilloscope images, self-test, and store download characters. The one we used was an Epson FX-850 (Fig. 1-7), which prints 290 characters/second at 10 cpi. Initial setup is a little tedious, but once made ready, this unit can operate in either single-sheet or continuous-feed modes by shifting the positions of a single lever and setting paper-mover sprockets. If you are setting up or adding to electronic test gear, both photos and linecut-type traces, especially for enduring records, could prove extremely valuable.

Another example would be Hewlett-Packard's 7090A DC3-kHz bandwidth and 33.3 kilosamples/sec recorder/plotter that combines waveform recording, digital plotting, analog recording, and automated measurements. A rather expensive system, this unit offers simultaneous sampling on three channels with 12-bit resolution, 1 000-word memory, six trigger modes, with all panel functions programmable, X-Y dumps, digital graphics, date and times, and it picks up selected data points from memory. Like its peer competition, it draws axes and grids, thereby eliminating any need for special graph paper. If all these modes of operation are requirements, then the multithousand dollar price is justified. If not, then a few hundred dollars will buy a fairly good working printer with limited functions.

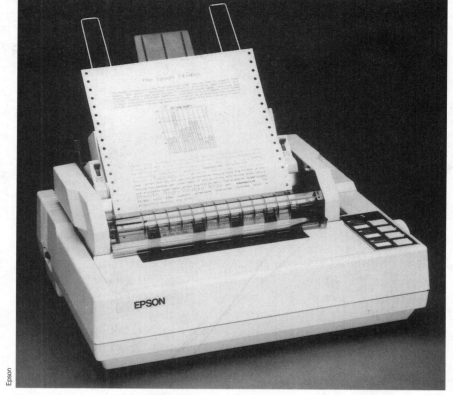

Epson

Fig. 1-7 *Epson's FX-850 9-pin regular-paper printer is also a source of line-cut printouts in this book.*

Any printer must be adaptable to whatever format the initial scope or analyzer has installed. An IBM format, for example, is quite different from our Epson FX series and you'll have to be very careful when using one versus the other. Fortunately, pinouts and cabled-contacts are distinct and the printer and its information source can only operate with the appropriate umbilical cords. Reading or readout accuracies and the actual means of creating an image are always most important. Some use ink pens; others use sensitized paper for thermal impressions; and another group offers ordinary ribbons for single-head printing (similar to a typewriter). Here, your author has no special advice—the choice is yours. Print images on ordinary paper, however, will certainly outlast those burned on thermals. Once again, need, applications, and cost are the three deciding factors, and sometimes they're hard to decide! Be patient, however, and evaluate the field, because prices on these electromechanical devices are not particularly firm and competition solves many dilemmas—especially as more instruments with printer readouts appear on the market.

This discussion could continue at considerable length over the respective merits and demerits of film versus print, but in the end, requirements dictate a hard choice, regardless of price or personal preference. For now, you pay for the film or pay for the printer. For semipermanent records, if kept away from light, a great deal is to be said for either or both; just be very cognizant of your instrumentation and objectives and use whatever delivers the best results.

In retrospect

This introductory chapter probably contains as much as should be developed with general, random topics. Before we're through, much more will be covered in varying detail, depending on available information and instrument innovations. What happened last year and the year before makes a great background, but today's events and timeliness of forthcoming instrumentation coupled with electronic applications and examples captured on printer paper or Polaroid film readouts, will supply prime information. In this way, we can helpfully demonstrate the special abilities and appeal of these instruments as they supply measurement solutions to many exacting tasks.

You should also understand how such equipment operates and what to expect of performance in ordinary and special circumstances, because it's difficult to work with equipment that's improperly or poorly understood. Consequently, although we can't offer actual factory training, forthcoming descriptions and working illustrations should serve as worthwhile aids in developing basic skills and certain innovative ones as well.

Although this book is by no means a television treatise, some of the latest developments in HDTV will be included as a technology update for general understanding of both systems and problems that are evolving in the 1990s. HDTV, for sure, will gradually have a considerable impact on most forms of over-the-air video communications and, therefore, is of more than casual importance and concern.

❖ 2
Waveforms and their analysis

WAVEFORMS THAT ORIGINATE FROM ELECTRON FLOW ALL COME from the passage of current through various active or passive devices that today are mostly conductors or semiconductors. Given the presence of sufficient operating voltage, the greater the excess current flow through either an active or passive electron carrier or multiplier, the less the voltage output, because more and more voltage is dropped (developed) across that device. Obviously, circuit-loading transistor/tube gain, and parameter biasing all play their respective parts but, generally, the foregoing statement is valid, and we will continue to discuss these items as the chapter develops.

A waveform, then, becomes nothing more than an electrical expression of phase, frequency, and magnitude that usually transports some specific intelligence. In passing through RF amplifiers and modulators, such intelligence must be detected (demodulated), the carrier suppressed or shunted, and finite information delivered to the end user, usually either an audio amplifier or cathode ray tube, including all the processing electronics lying in between.

If this information appears as baseband—that is, purely unmodulated with all its original components, minus carrier—then it is ready to be used in some appropriate system with only the addition of amplifiers, dividers, some phase shifting or doubling, and on into the final audio or video readout. AM and FM, of course, can be modulated onto an omnicarrier, but they must be demodulated separately because they are very different mediums of electronic transmission. Digital intelligence can also be included, and one day in the future, probably most or all telephony,

TV, and audio will be digitized and transmitted with much greater fidelity than is possible today.

All this, at the moment, is under active development and could appear more quickly than we suspect. Industrial and consumer products of all varieties would benefit considerably, especially those with more than their share of internal noise and transient problems. With the advent and widespread use of remarkably wideband fiberoptics, picture phones, and high-fidelity audio long-distance transmissions are very close to becoming realities in both the realm of digital and analog electronics. Certainly satellite broadcasting in the 4, 6, and 12 GHz bands could have noteworthy effects on sophisticated scrambling techniques that would also provide a great deal more in variety and fidelity of worthwhile programming. So you see, the user must know many disciplines of waveform development and processing to be able to interpret his or her findings when measurements are taken. That's why we have included this chapter as an aid to better understanding of the information you seek.

As waveforms develop

For now, however, let's exchange crystal balls for reality, and get down to business. Regretfully, we have to commence with sine waves for starters but, by including a little current along with power, the usual rote of peak (zero reference to pulse top), peak-to-peak (usually related to sine wave that swings positive and negative about some zero or dc axis), and just a pure dc level can be largely avoided. You should know, however, that periodic non-sinusoidal voltages or currents include basic sine waves and possibly their harmonics, constituting whole multiples of the original. We're speaking primarily of pulses and square waves, which will be considered shortly.

Figure 2-1 establishes the basic thought for the chapter. In it you see current and voltage (E_m, I_m) at somewhat different amplitudes, but in phase, with time progressing from left to right. Below, you see the two combined as power. Now, nothing is new about a sine wave. As usual, it alternates about some common axis, its positive alternation is equal and opposite that of its negative alternation. Sine-wave current does the same. When you combine the two, you're not talking about p-p power, but the work that current and voltage will do (the heating effect) that's equivalent to dc current and voltage. Therefore, the average (rms

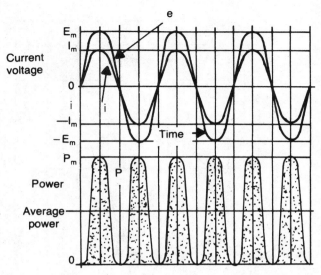

Fig. 2-1 *In purely resistive circuits, current and voltages are always in phase, and power is the product of rms E × I.*

or effective) power equals maximum current and voltage divided by two, and the actual power really means that the $\sqrt{2}$ root (mean square) values are multiplied together so that:

$$\text{Apparent Power} = E_m I_m/2, \text{ or } P_t = E \times I \text{ rms}$$

This power, then, shown in phase as average power in the illustration, will be considered rms power $P_t = I_m/\sqrt{2} \times E_m/\sqrt{2}$ also amounts to the same thing.

In circuits that have both $E \times I$ apparent power, and those with reactive elements (Fig. 2-2), such as capacitors and inductors, true power then becomes:

$$P_t = I^2Z \text{ or } E^2/Z$$

Taking into account, what one or more reactive elements are doing, we can easily deduce from these two simple equations that a power factor could be involved.

$P_f = I^2R/I^2Z$, which then becomes a ratio of R/Z, because the I^2 terms cancel. It further develops that $R = Z \cos \theta$ and $P_f = \cos \theta$. Consequently, the power in any reactive circuit equals:

$$P = EI \cos \theta$$

To prove this equation, just draw a simple right triangle with R on the adjacent bottom, X_L opposite, and Z as the hypotenuse.

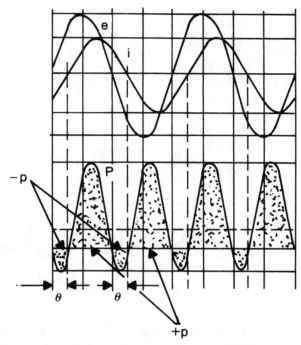

Fig. 2-2 *When reactive elements are introduced, impedance (Z) becomes a factor and currents and voltages are no longer in phase.*

Fundamental trigonometry states that the cosine divides the adjacent side by the hypotenuse, which, in this instance, is X_L/Z, and that proves the point that reactive power is obtained by multiplying maximum power by the cosine of theta.

Naturally, voltage and current are out of phase in any reactive circuit, and the X_L and R illustration we've chosen will bear this out. However, in calculating the impedance (Z), we find that:

$$Z = \sqrt{R^2 + X_L^2}$$

Then, the entire expression must be individually squared, added, and then the square root taken to find an equation that also takes into account frequency in the form of ω (omega) or $2\pi f$.

The square wave

A sinusoidal current or voltage often develops into what we readily recognize as a rectangular or square wave. That is, its duty cycle (the on/off times) amount to 50% high (a logic off condition) and 50% low (a logic on condition). Whenever any logic-operation voltage is low, as in either gates or flip-flops, device conduction occurs and current flows. In a high (off) condition, all conduction ceases and the waveform rises very close to the power supply, which, in semiconductor terminology, is commonly identified as V_{cc} or V_{dd}.

Known also as a *symmetrical display*, alternations that constitute such a voltage originate directly from the sine wave. This, for instance, can be graphically displayed in Fig. 2-3. Here you see the square wave itself at the top and then its development. The three ensuing drawings illustrate the addition of the 3rd, 5th, and 7th harmonics. As each additional harmonic is added, waveform tops flatten and risetimes and corners become sharper. If twelve odd harmonics were added, the rectangular wave would square out indeed and an excellent 50% duty cycle would be established. In dealing with high bit-rate logic, some of these risetimes must approach low nanoseconds and possibly even picoseconds to prevent false triggering among the logic chains, their buffers, memories, and arithmetic logic units.

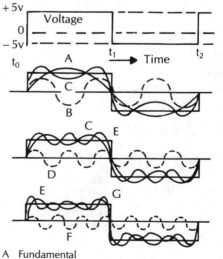

Fig. 2-3 *Any square wave consists of many odd harmonics in addition to the fundamental.*

A Fundamental
B 3rd harmonic
C Fundamental plus 3rd harmonic
D 5th harmonic
E Fundamental plus 3rd and 5th harmonic
F 7th harmonic
G Fundamental plus 3rd, 5th, and 7th harmonic

If interruptions or "glitches" were among such on and off voltage swings, their pulse widths and amplitudes might be just enough to false-trigger other words and bytes of information downstream, producing unintelligible readouts.

Because rectangular waves (square or otherwise) can be generated either by the analog accumulation of sine waves and their harmonics or digital switching, they all retain the same harmonic content, and this is what makes them so useful.

The top of any pulse or square wave contains intelligence required for low-frequency analysis, and the sides of these voltages supply high-frequency information. Between the two, any linear amplifier (analog) or digital system (1s and 0s) can be thoroughly tested for frequency response, linearity, and I/O (input/output) handling abilities. Naturally, you must remain within, or nearly within, the original specifications of the device under test. Then, barring stage, logic, or circuit faults, the proper response for the unit can be developed very quickly and accurately, presenting definitive good/bad readouts.

Where most problems arise are those in-between conditions where waveform tilt, a glitch or two, slight hum or ripple are present and might or might not have some specific effect on the final system or subsystem. We'll explain these conditions so that the general faults that might arise will not come as a complete surprise. It's impossible, of course, to foresee and discuss every single possibility or even probability, but certainly the major difficulties can be scrutinized and some general observations and solutions offered.

Waveshaping

A sine wave depends on the instantaneous value of E or I (voltage or current), coupled with additions of capacitors and inductors. Time constants immediately appear and waveshaping commences. In an LR circuit, for instance, time constant values are found by dividing inductance by resistance. Among RC circuits, $T_c = RC$, and nothing more. But, to find the instantaneous value of current and voltage in these circuits, you must use exponential equations, which also include base e (2.718) natural logarithms and a few tricky kinks that involve good old scientific notation, such as 3^{-2}, 10^6, etc. Furthermore, when e^{-1} appears, the table of hyperbolic functions says that this equals 0.367 9.

Hyperbolic functions are cousins of trigonometry's sine and

cosine, as well as its tangent, cotangent, secant, and cosecant. Written sinh, cosh, tanh, coth, sech, and csech, they directly involve e^x and e^{-x} and can be divided by or into the number "2" when calculating sinh, cosh, or sech and csech, respectively.

We're not involved in a treatise on mathematics in this chapter. I simply want you to know these engineering calculations really exist and why they're used. For instance:

$$\sinh x = \frac{e^x - e^{-x}}{2}$$

might not mean anything overwhelming at the moment, but it will, because hyperbolics are all part of the process of integration, just as peaked waves result from differentiation. These two very important processes are what waveshaping is really all about; repetition rates, time constants, waveform symmetry, and the introduction or removal of various circuit elements have major roles in developing all sorts of signals required to both transport and produce necessary intelligence.

Without going into either hyperbolic trigonometry or Euler's formula, which involves infinite series, let's turn to practical applications of what occurs in initial waveshaping as a result of both integration and differentiation. You might correctly associate this RC process with calculus, where differentiation measures the rate of change of variable quantities. Integral calculus is the inverse of differentiation and, given rates of change and instantaneous quantities, it can determine the value of a certain quantity at any instant. Said another way, differentials are simply stated or measured differences between successive values, and integrals are the sum or added successive values. Areas under curves, for instance, are found by integration, but the tangential slope of the curve is revealed by differentiation.

A very simple hardware illustration (Fig. 2-4) shows what is generally meant by both integration and differentiation. Of course, such RC couplings can be made vastly more complex,

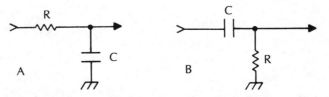

Fig. 2-4   *A simple example of (A) integration and (B) differentiation.*

and are also basic examples of low- and high-pass filters. Capacitors, you should recall, are most reactive at lower frequencies, but inductors are most reactive at higher frequencies. So, what happens when these RC networks are excited into action? Taking our previously formed square wave as an example, a great deal occurs (Fig. 2-5).

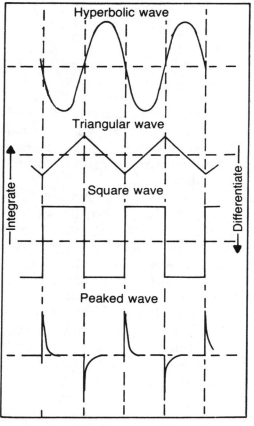

Fig. 2-5 *Integrated and differentiated square waves do interesting things.*

In the differentiation process, rapid rises and falls in resistor-developed voltages result in square-wave peaking on each half cycle of rise- and falltimes, provided that the capacitance is large enough not to unduly affect waveform symmetry, other than to form a suitable time constant with its resistor. The fast capacitor risetime on the leading edge precisely inverts itself on the square wave's downswing, and produces a negative-going equal and opposite reaction. As the square wave repeats, naturally, this differentiation process continues for as long as the train of pulses is generated. A short time constant, you might be interested to know, is when the RC time constant is equal or less than $1/10$ of the exciting voltage, and a long time constant amounts to $10 \times$ that of the applied voltage.

Differentiated voltages are primarily used as trigger mechanisms because of their short time durations and fast rises in almost any electronic equipment that uses clocks or sync circuits. Rarely do such trigger pulses fail to do their job and most are highly reliable in all sorts of timing applications. When and where the trigger points for such pulses occur, of course, is of prime importance and differences between ''successive values'' are highly significant.

Integrating potentials, on the other hand, are not at all suitable for precision triggering, but they can be used in ''build-up'' conditions where gradual rise- and falltimes or gentle slopes form driving circuits or troubleshooting situations. The waveshape at the top of Fig. 2-5 is our hyperbolic waveform, and it looks like nothing more than a sine wave with rounded top and bottom excursions, which it is. Then, by combining rectangular and hyperbolics, the resultant emerges as a triangular configuration, useful in ferreting out low-frequency distortion, high-frequency rolloff, clipping, and larger circuit or stage gain changes. In summary, at this point, integrating time constants are usually quite long, but differentiating time constants are rather short. As illustrated in Fig. 2-4, the former output develops across a capacitor; the latter appears across a resistor.

Interestingly enough, combination resistor and inductor circuits produce practically the same voltages and outputs as do RC circuits. In fact, the same general curves, time constants, and exponential factors are able to determine various values of current and voltage as before, remembering, however, that T_c for inductors becomes L/R, instead of RC for capacitors. Remember that dc coil resistances are comparatively negligible, but those of capacitors are often infinite. So, don't confuse the two where dc voltages of any magnitude are involved, unless you're deliberately looking for either choke or peaking effects versus coupling or bypasses.

While on this particular subject, you should know that differentiating or integrating a periodic sine wave induces only amplitude and/or phase change, but it does not disturb the actual waveshape. So, pure sine waves are always called *fundamentals*—they don't change shapes, even though passive-element capacitors and inductors exist in the circuit.

Pulses help too

Using rectangular waves in appropriate analog circuits, such as linear amplifiers and digital clocks or drivers (Fig. 2-6), can produce some interesting effects, all of which identify a variety of important problems. Preshoot and overshoot, for instance, is often actually used in analog video circuits to offer accentuated transitional change from black to white, or their appearance simply signifies some nonlinear circuit defect. High and low frequency losses are easily identified by their respective corner rounding or sloppy falltimes. Ringing and hum are quite similar, except that hum shows in both high and low transitions, and ringing occurs primarily during amplifier cutoff. Where amplifiers actually lose amplitude in the top-bottom low-frequency portions of the waveform, this applies to fundamentals only, and has nothing to do with accumulated harmonics.

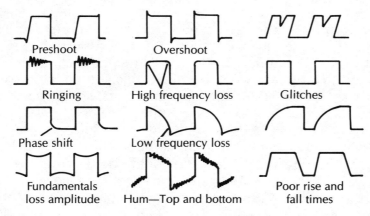

Fig. 2-6 *Square waves usually tell the problem story in both analog and digital circuits.*

In digital applications, rise- and falltimes and traditional ''glitches'' are prime fault indicators, plus pulse stretching, repetition rate failures, and various set and reset operations that are always present in digital logic. Pulse circuits of course, differ only from square waves in their pulse widths, and might even have the same repetition rates and periods (Fig. 2-7). Leading edges are usually critical because they establish timing or triggering for parts of a large segment of the electronic system. If the entire pulse is used for sync or period operations, trailing edges

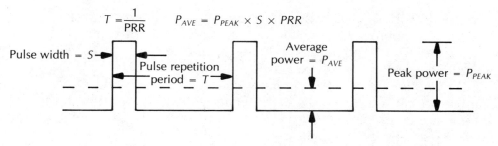

Fig. 2-7 *Repetitive pulses, showing width, duration, and power.*

must maintain their sharpness also, and pulse widths cannot be allowed to vary if they are normally fixed and repetitive.

Should this repetition rate amount to, say, 2.5 MHz, then its period would be $1/2.5 \times 10^6$, or 400 nanoseconds (400×10^{-9}), and that's pretty fast. Of course power also enters into any pulse evaluation, and you will find that:

$$P(average) = P_{peak} \times pulse\ width \times pulse\ repetition\ rates$$
$$or\ P_{average} = P_{peak} \times S \times PRR$$

Therefore, pulses with longer durations, large amplitudes, and high repetition rates require considerable power and must be handled by test instruments with care, whether contact comes in troubleshooting or simply measuring to establish adequate operation. Oscilloscope probes, you recall, can only handle so much current and voltage; after that, internal resistors and even capacitors simply turn red and melt away. Peak power here is the actual pulse power, and average power includes both pulse on and off times. So, to find the duty cycle of each pulse, divide pulse repetition period by the pulse width and:

$$Duty\ cycle = PRR/S$$

Coupling circuits

Let's deviate just a little here and dwell for a time on conductive and nonconductive combinations that you'll find primarily in low-frequency circuits. Don't be surprised if some of them look just like low-pass and high-pass filters or even a high impedance trap. Simple value changes, along with large increases in frequency, do remarkable things to basic RLC passive components. In any reactive LC condition, ωL and $1/\omega C$ must always be taken into consideration, especially where inductors peak and capacitors filter. Furthermore, resistors in reactive circuits also play a part in amplitude and phase changes, and so they must be regarded as part of any RL, RC, or RLC network.

As you see in Fig. 2-8, we have what are called *conductive configurations* and those that are *nonconductive*, in addition to combinations of the two. All these small networks simply transfer electrical energy from one series of amplifiers or oscillators to another, matching impedances and/or filtering transients along the way. There are also RC and LC couplings (Fig. 2-9), which are

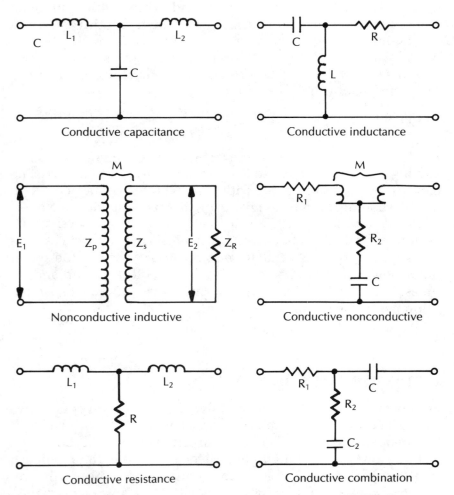

Fig. 2-8 *Conductive and non-conductive circuits in basically low frequency applications.*

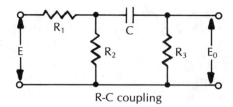

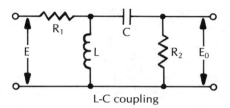

Fig. 2-9 *RC and LC couplings are also available.*

often used under specific circumstances where voltage division and dc blocking are important, or where ac voltages become significant and little dc remains in the circuit. Remember, too, that in predominantly capacitative circuits, current always leads voltage, and in predominantly inductive circuits, voltage leads current. Therefore, any coupling circuit must be chosen on the basis of phase shift, amplitude and power loss, and of the respective I/O impedances. With both reactances and resistances entering the equations, plus the addition of phase shifting (operator J) expressions, these networks become somewhat "hairy," and we had best not become involved mainly because of their length and confusing terminologies even though the associated math extends beyond algebra.

In the nonconductive/inductive condition (Fig. 2-8), you see transformer coupling, Z_p stands for primary impedance and Z_s represents secondary impedance. Here, the equations are quite simple, so that in any ideal transformer:

$$\frac{E_2}{E_1} = \frac{N_2}{N_1} = \frac{I_1}{I_2}$$

$$\text{and} \quad \frac{Z_{in}}{Z_{out}} = \left(\frac{N_1}{N_2}\right)^2$$

$$\text{So} \ Z_p = \frac{E_1}{E_2} \times Z_2$$

Be careful, however, when assessing operational parameters

of any transformer. Power out has to be less than power in. Further, recall that center-tapped transformers signify polarity changes, and multitapped transformers are usually used for secondary load matching or secondary voltage sources. In addition, should primary and secondary windings be of opposite polarities, then you can expect a 180° phase reversal to appear across the secondary. *Maximum transformer efficiency* is reached when copper losses equal those of iron losses, and *efficiency* is measured as the ratio between input and output power plus such losses. The percentage of voltage regulation, then, becomes the difference between fully loaded and unloaded secondary voltages divided by full load voltage and multiplied by 100. In signal transformers, objectives are usually coupling for best waveform symmetry, but in power transformers and current drivers, plain rms $E \times I_{power}$ is the prime requisite.

Filters

Returning to filters for a moment (especially those in simple power supplies), there are usually two types: capacitative input or inductive input (Fig. 2-10). More sophisticated arrangements have active filters, where capacitance is multiplied by the gain of

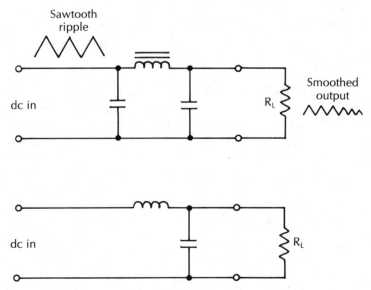

Fig. 2-10 *Ordinary capacitor and inductor power supply passive ripple filters.*

an associated transistor and it becomes highly effective, especially in half-wave circuits.

Look at two common types for basic familiarity, regardless of whether the inputs are half wave, full wave, or bridge types, because the end result amounts to a smoothing of dc current ripple for system operating voltage. Nor are we concerned with either current limiting or voltage regulation at the moment—just what a power supply filter will or should do at 60 or 120 Hz, which means full- or half-wave rectification.

Illustrations of both types are shown in Fig. 2-10. Although an inductance input filter maintains voltage at similar value as the capacitance filter when rectifiers are not conducting, voltage across the inductor drops rapidly during conduction even though it remains fairly constant thereafter. The more common capacitance input type has capacitors shunting both input and load, with a series inductor (or resistor) acting as an ac/dc choke between. During conduction, the capacitors store energy and deliver this energy back to the load whenever voltage tends to decrease. The inductor opposes changes in current, offering double action in controlling dc ripple. As you see in the waveform about the capacitor input illustration, a sawtooth is formed as a result of rectifier conduction, whose amplitude depends on incoming voltage and whose frequency will double with full-wave detection. Regardless, an input sawtooth must always be there, or you have a leaky or open input capacitor or a bad rectifier. If either input or output capacitor is defective, a drop in output voltage, accompanied by increased ripple is ensured.

As said previously, some coupling and filter circuits do look alike. When you compare Fig. 2-11 with Fig. 2-8, you'll recognize several similarities. These filters, of course, consist principally of inductors and capacitors, arranged in series or parallel, and are actually tuned circuits to either pass or obstruct certain frequencies. Their makeup can be very simple, as illustrated, or highly complex, including delay lines, parallel and series-tuned traps, and slot or bandpass arrangements that permit only certain segments of intelligence to proceed from one point to another.

During passage of low-frequency information, the filter in Fig. 2-11A offers little opposition through its inductors, and the reactance of the shunt capacitor is high. However, as these frequencies rise, both inductors increase markedly in reactance, while the shunt capacitor's reactance is low. Reverse these conditions and you have the LC filter shown in Fig. 2-11B. Here, you

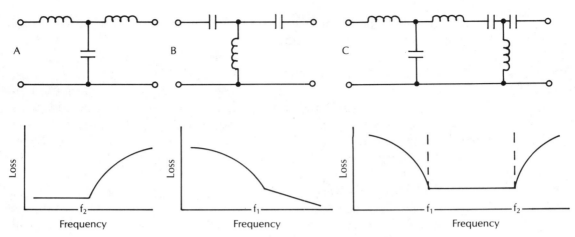

Fig. 2-11 *Fundamental lowpass (A), highpass (B), and bandpass filters (C).*

see two capacitors connected in series and between them is an inductive shunt. As input frequencies decrease, the capacitors cease to behave like short circuits and increase their opposition, while the inductor's reactance soon becomes low, and passage of any information in that spectrum is minimal or nonexistent.

With multiple inductors and capacitors combined (Fig. 2-11C), low and high frequencies are attenuated proportionally to whatever intelligence needs processing, and only some band of center frequencies is permitted through. Illustrative curves showing lows with respect to frequency give you an idea of such filter performance; the sharper the cutoff, the more desirable the filter. At RF frequencies in many video consumer products, surface-wave acoustical filters (SWIFS) now largely supplant such conventional discrete device filters and do provide markedly sharper band skirts and much better adjacent channel rejection, while maintaining adequate signal-to-noise ratios (S/N). This latter condition can often be overlooked by cheap product designers, thus producing audio hissing and annoying flecks of video "snow," either of which is unacceptable, except under very weak input signal conditions. Ceramic filters are also becoming more popular—these are further improvements over older types that have been in common use for years.

Before leaving the subject of coupling, let's take a final look at what's called a *bandstop filter*. You should recall that series resonant circuits permit maximum current flow and little opposition to frequencies where L and C cancel. Parallel resonant circuits supply minimum current and maximum impedance to these same RC combinations. Should you have double sidebands involved and one required elimination, you could proceed, as shown in Fig. 2-12.

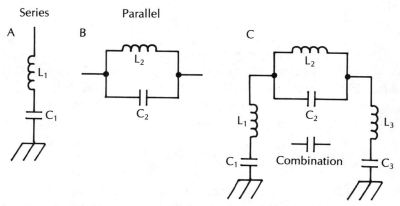

Fig. 2-12 *Bandpass filters consist of parallel and series traps combined.*

Put the series and parallel circuits together, as shown in Fig. 2-11C, and you have an effective bandstop of whatever frequency you wish to remove. Here, the L_1C_1 series combination shunts this unwanted voltage to ground, then the C_2L_2 parallel-tuned pair offers high impedance to this same frequency, and L_3C_3, tuned the same as L_1C_1, removes any remainder. This circuit is also called a π *trap*, which still enjoys common usage in various forms and inductive-capacitative values.

As a roundup, high-pass filters pass high and reject low frequencies; low-pass filters pass low and reject high frequencies; and bandpass filters reject both high and low frequencies and permit only center information to continue. Don't fail to remember, however, that in any arrangement of inductors, capacitors, and resistors, there are distinct phase shifts, often combined with loss of amplitude, so one effect has to be balanced against another when evaluating or designing any coupling or filter circuit.

Video

No chapter on waveforms could be complete without an in-depth study of the video signal process. In computer readouts, closed-circuit television, Teletext, CATV, satellite broadcasting, and the family's home television in its various faces is making the world a far more exciting and rewarding place to live. And as the '90s move toward the close of the century, we firmly expect more people everywhere to be working at home, shopping to a considerable extent over cable television, and having a home direct view or large-screen projection set with baseband monitor, audio/video inputs and possibly outputs, to become the prime source of all family entertainment. Moving picture theatres could all but vanish as satellite and cable programs reach a majority of all U.S. TV households, and a good TV college education will, undoubtedly, become readily available as master teachers are attracted and learn to use this tremendously useful and available medium.

The NTSC signal

Two things we won't do in this section are review the history of television and discuss black and white characteristics, nor will we bore you with inconsequential details. But, we will bring you quickly up to date with appropriate information about signals and then signal characteristics, illustrated with pertinent waveform applications. Because standard color TV parameters are now broadcast by virtually all stations, only those characteristics will be discussed.

Two fields are in every frame, 262.5 lines in each field, and 30 frames/second, which you view on your TV screen at a 59.94 Hz complete picture rate.

If you set the time base of your oscilloscope at 2 or 5 milliseconds/division, you will look at one or two fields of video and sync information. However, if the scope's time base is set for 10 or 20 microseconds, then you will see one or more of the 525 horizontal scanning lines, plus a horizontal sync pulse, and a color sync alternation on its back porch of some 8 cycles, called *color burst*. Both such waveforms are drawn for information in Fig. 2-13, with each being referenced to the traditional IRE (Institute of Radio Engineers) 1-volt video scale of 140 IRE. The latter is highly important since it has become an accepted measure of the video broadcast signal and is useful not only for studio setup and monitoring, but in evaluating receiver responses in all color-producing video products, either with or without video demodulators. Freed of modulators, naturally, baseband unmodulated information does not need to be demodulated, and will be used

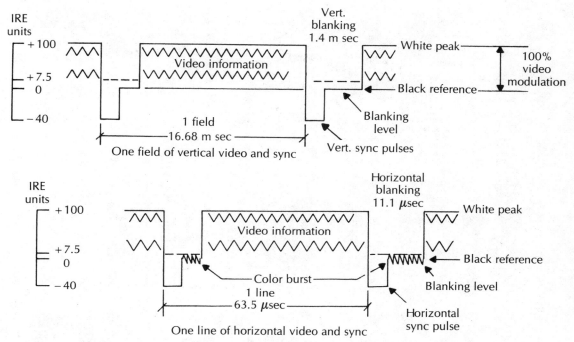

Fig. 2-13 *Single cycles of vertical and horizontal sync information referenced to IRE scales.*

more and more with any direct hookup video, which certainly
includes the vast number of pure monitors and TV/monitor combinations that are now coming into the market in large numbers.
For the sake of convention and continuity, we'll show these signals in negative polarity, but when you view illustrated NTSC
sync, a mental positive polarity flip will be needed to relate the
two to one another.

In Fig. 2-13, note that only four IRE references have been
used: – 40 for sync tips, 0 for blanking, + 7.5 for black reference,
and + 100 for peak white. If you remember these four and that
sync pulse amplitudes occupy 25 to 30 percent of the total waveform between – 40 and 100 IRE, then you already have a working
understanding of the fundamental video signal, and we can
readily fill in the rest.

Ordinarily it would be preferable to forego some of the extensive detail in the next illustration (Fig. 2-14), but because engineers will be using this book, such an urge had best be
suppressed in the interest of maximum accuracy. Instead, a general explanation will be offered of its content that's intended to be
both inclusive and incisively succinct.

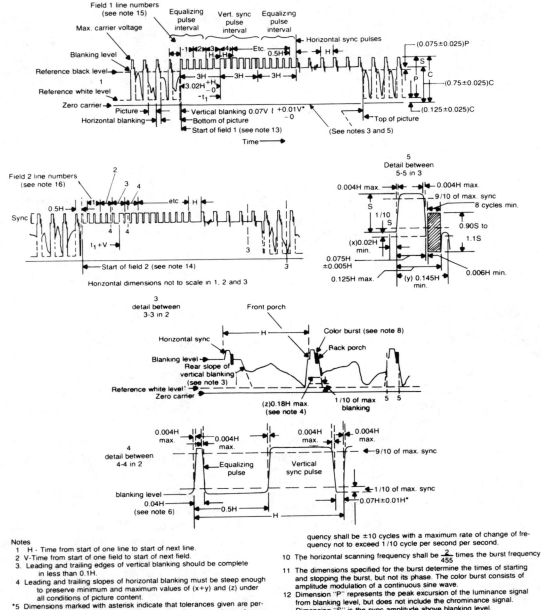

Fig. 2-14 *FCC specified sync and blanking parameters with explanatory notes.*

Notes

1 H - Time from start of one line to start of next line.
2 V-Time from start of one field to start of next field.
3 Leading and trailing edges of vertical blanking should be complete in less than 0.1H.
4 Leading and trailing slopes of horizontal blanking must be steep enough to preserve minimum and maximum values of (x+y) and (z) under all conditions of picture content.
*5 Dimensions marked with asterisk indicate that tolerances given are permitted only for long time variations and not for successive cycles.
6 Equalizing pulse area shall be between 0.45 and 0.5 of area of a horizontal sync pulse.
7 Color burst follows each horizontal pulse, but is omitted following the equalizing pulses and during the brood vertical pulses.
8 Color bursts to be omitted during monochrome transmission.
9 The burst frequency shall be 3.579545 mc. The tolerance on the frequency shall be ±10 cycles with a maximum rate of change of frequency not to exceed 1/10 cycle per second per second.
10 The horizontal scanning frequency shall be $\frac{2}{455}$ times the burst frequency.
11 The dimensions specified for the burst determine the times of starting and stopping the burst, but not its phase. The color burst consists of amplitude modulation of a continuous sine wave.
12 Dimension "P" represents the peak excursion of the luminance signal from blanking level, but does not include the chrominance signal.
   Dimension "S" is the sync amplitude above blanking level.
   Dimension "C" is the peak carrier amplitude.
13 Start of Field 1 is defined by a whole line between first equalizing pulse and preceding H sync pulses.
14 Start of Field 2 is defined by a half line between first equalizing pulse and preceding H sync pulses.
15 Field 1 line numbers start with first equalizing pulse in Field 1.
16 Field 2 line numbers start with second equalizing pulse in Field 2.

As you were forewarned, sync pulses and blanking levels are now on top (reversed), while video occupies the space between zero carrier and reference black level. At the top are four sync pulses, along with video, trailed by six equalizing pulses in the time of three horizontal lines, then six vertical sync pulses, followed by another six equalizing pulses (all in the space of six horizontal lines), six horizontal sync pulses, and another three pulses with video. This completes field 1, and sync is now ready for the introduction of field 2 (two fields compose a frame).

Once again, pulses with video begin the field, followed by six equalizing pulses, six more vertical sync pulses, six equalizing pulses, and more horizontal pulses to finish the sequence.

Details for both the equalizing and vertical pulses are shown in Fig. 2-14 at the bottom. At the top right, a horizontal pulse with burst on its back porch is drawn; and a pair of horizontal line cycles with both sync and burst appears in part three, along with explanatory references in the notes to the right.

Notice that the line and field starts, along with their dimensions, are symbolized by the letters H and V. Specifications for equalizing pulses are between 0.45 and 0.5 (half) the area of any horizontal sync pulse, and burst frequency, set at 3.579 545 MHz, has a tolerance of only ± 10 Hz. Observe also that horizontal frequency is specified as 2/455 of burst—a convenient reference.

Virtually all measurements for sync are tied to burst or horizontal lines, and this serves well in both transmitter and receiver countdown circuits that automatically control sync throughout the broadcast day. Pulse durations and risetimes, of course, are critical and must be maintained for adequate transmission and reception.

When vertical sync and equalizing pulses are distorted or missing in either broadcast transmissions or reception, vertical rolls, jitter, and other unpleasant symptoms are apparent to the viewer, because the absence of sync lock prevents a stable 4:3 (aspect ratio) raster. With horizontal sync problems, instability and picture tearing are the obvious faults. A good example of these would be some of the scrambled pay-TV programs available over the air or cable, some of which also produce negative pictures.

Test signals

Obviously, vertical and horizontal test signals are available in even ordinary gated rainbow color-bar generators to check raster (blank picture) linearity and effectiveness of countdown ICs and discrete horizontal/vertical hold controls, if there are any. However, the most critical portion of any transmitter/receiver is its luminance and chroma functions. Here, highly sophisticated test equipment is required to do a thorough job. Once again, Tektronix and other suppliers with their excellent waveform generators offer superb examples of what to use and how in delivering test signals to and through both transmitters and receivers. Although the number of equipment suppliers is growing, many more manufacturers of video products, especially HDTV proponents, will appear in the 1990s. The same holds true for satellite earth station test equipment, which requires signal generation between 4 and 12 GHz for the vast variety of fixed and direct broadcast satellites now or about to be in synchronous orbit above the equator. Even now, preliminary 24 GHz testing is underway for additional space "birds" of the future.

Some of the more useful test signals generated by available Tektronix test equipment are shown in Fig. 2-15; they are actual oscilloscope photos. In the upper left corner are the six cycles of multiburst, preceded by a 100 IRE peak bar, along with horizontal sync pulses and burst. Any tilt in the horizontal pulse means low frequency rolloff, and multiburst oscillation distortion or adjustments in amplitude signify problems between the 0.5- and 4.1-MHz cycles of oscillation. The top right waveform amounts to a horizontal pulse and burst on its back porch, a 6-step modulated staircase, 2T unmodulated and 12.5T modulated spike of voltage, and a rectangular pulse. These are used, respectively, for checking nonlinearities and differential phase errors, luminance gain, chroma gain and phase lead or lag, and any rounding or ringing that might be apparent in tuned or video amplifiers.

At the bottom left are the six white, yellow, cyan, green, magenta, red, blue, and black color and noncolor bars (less I and Q signals) that are generated by any NTSC color-bar generator. Notice that their top amplitudes begin at + 100 IRE and conclude at about—18 IRE. Although these are quite useful in the studio and on monitors for adjusting proper transmitter output, their adjustable levels in consumer products make them less desirable than the 3.56-MHz gated rainbow patterns—all of which are described at length in chapter 4 on vectors. On the bottom right is VIRS, the *vertical interval* (color) *reference signal*. Its chroma envelope extends from + 50 to + 90 IRE, along with luminance reference, while black level appears at + 7.5 IRE, just above hori-

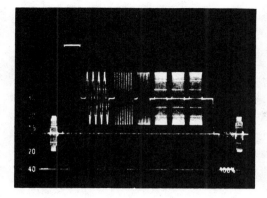

Multiburst Per FCC § 73.699

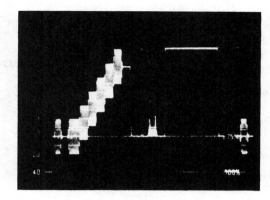

Composite Test Signal Per FCC § 73.699

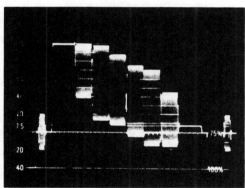

Color Bars Per FCC § 73.699

Vertical interval reference signal (VIRS).

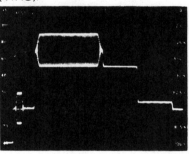

Fig. 2-15   *The four most widely used test signals for both transmitters and many video receivers.* Tektronix

zontal blanking. Once more, this is primarily a studio setup device at the moment, but as additional receivers and monitors are built with better and more accurate comb filters and video amplifiers, such waveforms will become extremely useful. In this pattern, chroma must be in phase with burst for color circuits to operate properly. If either burst or chroma is out of phase, the image hue develops the usual purple or green people. Some receivers (such as those built by General Electric) have VIR-correcting circuits that take care of these annoyances without distorting the chroma signal. Others have "idiot buttons" with pretuned arrangements that deliberately spread flesh tint hues from 90° (quadrature) to plus 105° to 120° and distort greens and/or blues as well.

The gated rainbow generator is especially useful because its reproduced 10 bars are gated at 30° intervals, extending from yellow and orange through red, magenta, blue, cyan, and green. If these bars are not improperly blanked, they are useful in 10-petal vector applications that will easily show alignment, jitter, lack of color burst, critical R-Y and B-Y phase angles, red and blue amplitudes, and other normalities or abnormalities in color receiver circuits. The next chapter should give you considerably more insight into the operation and application of both NTSC and gated rainbow vectors.

VITS and VIRS

Before we leave this chapter, perhaps it would be effective to illustrate both VITS and VIRS in artist-drawn waveforms so you can realize the full significance of these two vital signals.

The VITS signal (Fig. 2-16) occupies two lines of the vertical retrace (blanking) interval. On line 17, field 1, you see the familiar six cycles of multiburst, followed by the traditional set of colors bars in field 2. On line 18, are composite signals that consist

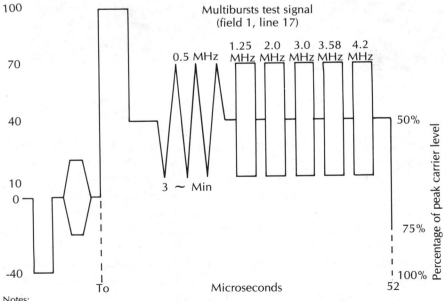

Notes:
1. A breezeway, as shown between bursts, is recommended. Each burst = 60 IRE units peak to peak.
2. To = Nominal start of active portion of line 17, field 1
3. Rise and fall of white bar shall have risetime of not less than 0.2 μS.

Fig. 2-16   *The valuable multiburst portion of VITS.*

of staircase and 2T, 12½T sin² pulses and an 18-microsecond window which have already been described.

Multiburst begins with three cycles of 0.5-MHz oscillation followed by 1.25-, 2-, 3-, 3.58-, and 4.2-MHz separate oscillations. With these you can establish what any series of amplifiers and demodulators are doing, their bandpass, and high- or low-frequency rolloff characteristics. Even in evaluating video discs, video cassettes, and standard TV receivers, this multiburst pattern is invaluable to establish what the luminance circuits are producing in the way of high-frequency resolution or low-frequency gain and loss.

VIRS (Fig. 2-17) occupies both fields 1 and 2 of line 19 of the vertical blanking interval. At the top of the figure is the chrominance reference bar, just a few microseconds to the right and above the usual nine cycles of color burst. Color burst lies between – 20 and + 20 of the IRE measure, and chroma reference occupies 40 units between 50 and 90 IRE. As long as burst and chroma are in phase, and amplitudes of all other parts of VIRS are the same as those illustrated, the vertical interval (color) reference signal signifies adequate chroma response throughout tested amplifiers. However, when burst and chroma references are either out of phase or reduced significantly in amplitude, color levels and phase changes so that color distortion is acutely visible on the viewing screen.

Full fields are graphically illustrated in Fig. 2-18 by a picture taken directly from a television receiver at its wideband video detector. Here you see all six oscillations of multiburst, proving the broadcasting station and receiver are putting out and receiving 4-MHz bandwidth. The usual horizontal sync pulse follows with chroma burst on its back porch, then a modulated staircase, with both 2T and 12.5T luminance and chroma spikes of voltage following. Then, comes the rectangular window, another horizontal sync pulse and, finally, the line-19 chroma VIR signal. Observe that multiburst is adequate, but not full amplitude at 4.1 MHz. As you see, the staircase is reasonably linear and the ensuing 2T luminance voltage is at full amplitude. However, chroma is foreshortened and not at the top of the staircase; it remains somewhat down from the 50 and 90 IRE measurements specified.

Is this a passable waveform for both receiver and broadcast station? Yes indeed; it's considerably better than many you'll see. The camera, by the way, was a Tektronix C-5C, and the waveform time based at 0.1 msec plus a 5 × expander. Where older broad-

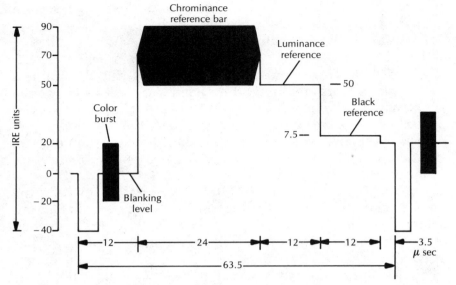

Fig. 2-17   *Line 19 of the vertical blanking interval displays VIRS. For proper color, chroma and burst must be in phase.*

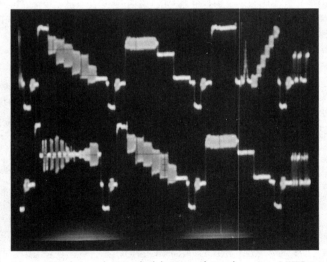

Fig. 2-18   *Lines 17 through 19, fields 1 and 2, showing VITS and VIRS as they appear (after demodulation) during TV broadcasts.*

cast stations use notch filters in their outputs, bandpasses are not always 4 MHz which, of course, affects receiver characteristics. If the whole truth was known, some of the top U.S. receivers are now capable of displaying better pictures than they often receive.

Poor film, tape, noise, black-level shifts, and broadcast misadjustments all contribute to transmissions that could certainly be improved over what is often modulated on the airways today. It will probably take a lot of excellent satellite reception to force terrestrial transmitters into making that extra effort. When they do, Americans are in for a rare treat. The NTSC system, authorized December 17, 1953, might not be better than Germany's PAL but, used to its fullest extent, presents an excellent picture that will only be surpassed by 708-1 125 line high resolution video from any and all geosynchronous satellites and many terrestrial systems that are due about 1995, along with screen-size aspect ratios that are increased from 4:3 to 16:9.

❖ 3
The extraordinary oscilloscope

FIRST USED ABOUT 1897, THE NOW-ESSENTIAL OSCILLOSCOPE began as a simple cathode ray tube, two sets of vertical and horizontal deflection plates and a rudimentary power supply (Fig. 3-1). Initially a very low frequency device with questionable accuracy, the normal analog service oscilloscope now delivers 2 to 5% accuracy on both X and Y axes, accommodates low mV to high V inputs, and features time bases that measure from seconds to nanoseconds (A $\times$ 10^1 to B $\times$ 10^{-9})—and even less with 10 $\times$ magnification.

Rigidly defined, this instrument exhibits results of electrical impulses on phosphors of some selected cathode ray tube. However, it was not until the advent of vacuum tubes and somewhat linear (an extension in one direction) amplifiers and the introduction of calibrated time bases that the oscilloscope really became useful. Now, instead of audio-type kilohertz frequencies and nontriggered time-base characteristics, these scopes, which are now almost universally solid state, extend passbands to gigahertz and time bases down to picoseconds (10^{-12}), if you have the need and want to pay the cost. All of this, coupled with more efficient cathode ray tubes, switching power supplies, extended linear operation and even BrightEye™ tubes, said to be over 1 000 times brighter than those on ordinary scopes, have all contributed to making the oscilloscope an essential instrument for almost any analog or digital design or repair-shop endeavor.

In addition to analog (signal processing) varieties are digital, sampling, storage, logic, various types of analyzers, time-domain reflectometry, and other highly specialized types that might or might not be in general circulation, depending on spe-

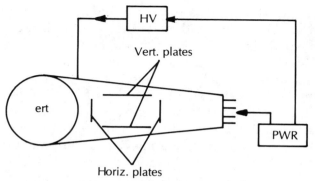

Fig. 3-1   *A simple oscilloscope has four plates, a cathode ray tube, and a power supply.*

cific uses. All prime types will be discussed in as much detail as the manufacturers permit, including block diagrams and even limited schematics. In this chapter, however, the lead topic is restricted to analog oscilloscopes, which are the ordinary work-horses of the commercial/consumer industry. The military and special projects we'll leave alone. However, many ordinary descriptions and applications apply equally to these latter group-ings, too, so they're generally included as well.

How they work

Once the V/H deflection plates, power supply, and cathode ray tube are developed, push-pull amplifiers and a time-base oscilla-tor are required to change the equipment from a passive, nonam-plified display into a high-sensitivity, wide-ranging instrument that is capable of displaying peak-to-peak, peak voltages and dc levels, as well as the times of each alternation, pulse, or aberra-tion. Notice that root-mean-square (rms) is specifically not included, because rms is the ac heating value that corresponds to dc current drive through a resistor. So, you must divide p-p by 2.828 or peak by 1.414 to establish a true rms reading. These readings are extremely important because rms values are becom-ing much more common in sine-wave descriptions and analysis. When evaluating pulses and square waves, the swing from com-mon to power supply (**rails**, as the Brits say), is read directly without further interpolation; that is, if you're using either a 10 × low-capacitance or direct probe. A video-detector probe, on the other hand, will demodulate a complex or sine-wave signal as it operates on the signal's envelope and it usually produces a half-wave current output that develops into voltage across its built-in load. This output is nonlinear, however and, unless very carefully designed, it will often produce circuit loading, diminished amplitudes, possible oscillations, and questionable waveform patterns. It is best to stick with the 10 × LC probes wherever you can and avoid obvious problems. A half-wave anything usually doesn't cut the mustard in our advanced state of electronics. As a result, the better consumer television equipment has universally adopted full-wave synchronous detectors that have virtually none of the failings that are associated with the germanium (0.2-V drop) half-wave group.

The illustration in Fig. 3-2 shows the three types of oscilloscope probes that are ordinarily available from electronic factories and distributors. The better LC probes usually contain both inductor and capacitor trimmers to compensate for both high- and low-frequency inadequacies that would result in false readings.

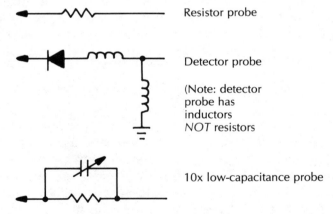

Resistor probe

Detector probe

(Note: detector probe has inductors NOT resistors)

10x low-capacitance probe

Fig. 3-2   *The three common types of passive probes are available for most oscilloscopes.*

Lesser-value probes only have variable capacitance compensation for low frequencies. Direct probes with only 10-kΩ resistors will often load any high-impedance circuit under investigation and even cause oscillations in the more sensitive systems—especially at higher frequencies. What you don't want, however, is a probe adjustment that looks more like a sine wave than a rectangle, such as the lower trace in Fig. 3-3. The upper trace isn't adjusted properly, either, because the sloping sides decidedly indicate poor

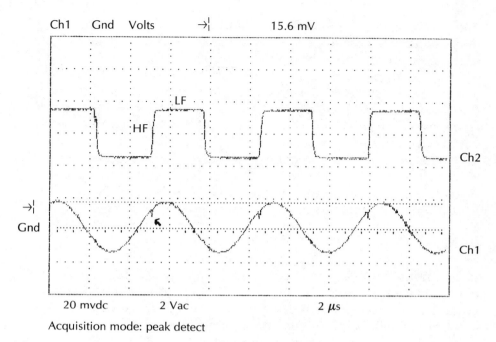

Ch1   Gnd   Volts   →| 15.6 mV

LF

HF

Ch2

→|
Gnd

Ch1

20 mvdc   2 Vac   2 μs

Acquisition mode: peak detect

Fig. 3-3 *Probe adjustments are always made on "perfect" rectangular waves and never on sine waves.*

risetimes as well as other high-frequency (HF) capacitance problems. Trace tops and bottoms are reasonable for low frequencies (LF). Just to make instrument life a little more interesting, an automatic measurement from ground to peak was requested of this oscilloscope and printed. Notice anything else? How about the glitch in the channel 1 sine wave? Is this a perfect waveshape? Of course not! Therefore, use the scope's own square wave to adjust the LC probes for at least the lower frequencies and a separate, high-quality square wave generator for high frequencies in MHz. You'll be glad you did! True, the LC probe changes your scope's nominal input impedance from 1 to 10 MΩ which also means you lose maximum scope sensitivity by a factor of 10. So, don't forget to mentally compensate for the additional 9 MΩ, if your oscilloscope doesn't already do so on its front panel.

The basic scope

The equipment selected for this discussion is a 60-MHz HM 604 oscilloscope that was manufactured in West Germany by Hameg and introduced to the U.S. just before 1991. A 3%-accurate, dual-trace (not dual-channel) measuring device, this scope is moderately priced, can evaluate signals between 1 mV and 20 V/ div., and features a time base in 23 calibrated steps from 50 nsec/ div. to 1 sec/div., along with a x10 magnifier, extending the high end to 5 nsec/div. Ideally suited for almost all types of bench and field services, the HM 604 also features an active TV line and frame sync separator, an after-delay trigger, 12-kV accelerating voltage, a component tester, an X-Y display, dual channel inverts, 1-kHz-to-1-MHz calibrators, LED indicators for triggering and off-screen vertical deflection, channel-switching holdoff, an adjustable internal delay line for expanded signals, and a P43/123 5″ cathode ray tube, in addition to a rear-panel BNC-coupled Z-input, and time-base, Y outputs. An optional lighted graticule is necessary.

Most of our foregoing discussion is outlined in a pair of illustrations from a sister scope that has only a few modest changes, such as dual X-Y inversion, TV sync separation, and control placements. Therefore, the two are described as one for the front panel, while signal processing is exact. The vertical amplifiers and their attenuators are first.

As shown in Fig. 3-4, all Y controls are neatly arranged for both amplifiers with ground, ac and dc inputs, alternate and chopped selections, single- or dual-trace displays, and inversion(s), in addition to a magnification pull and channel Add button (not shown). Probe inputs are BNC connectors that have common (return) contacts and separate ground jacks alongside. Attenuation is plainly marked in millivolts/centimeter, and Y-positioners move either trace to convenient positions. When one channel is engaged, internal sync is also applied from the same

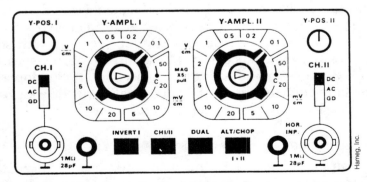

Fig. 3-4 *A front-panel example of the 60-MHz dual-trace HM 604 oscilloscope.*

channel. Dual channels can also be added or subtracted from one another and sensitivities can be increased to 1 mV/div. with the × 5 magnifier. In the alternate trigger mode, signals with different frequencies can be compared.

An actual schematic of the X/Y board appears as Fig. 3-5, which illustrates the two pairs of push-pull Y amplifiers and their RC compensators, followed by guide diodes above and special trigger preamplifiers below. With further amplification, signals of various descriptions continue toward the vertical plates of the cathode ray tube, with switching supplied by amplified triggering, which controls the various modes of operation. Each input channel has a coupling selector, a variable attenuator, and diode-protected FET inputs (not illustrated).

Before receiving signals, be sure that probes supplied with the equipment are adjusted precisely so that patterns (Fig. 3-6) are exact and neither round-off the square wave nor overshoot. As you can see, these are not ordinary test probes, because they are LC adjustable for high and low frequencies, and the HM 604 furnishes both 1-kHz and 1-MHz calibration facilities. An HZ51 probe, for instance, offers a risetime of less than 2 ns, a bandwidth of 150 MHz, 10 MΩ of input resistance, cable length of 1.2 m, and a maximum voltage accommodation of 600 volts. Use a probe passband double that of its accompanying oscilloscope.

Displayed signals can appear single channel—either channel 1 or channel 2. In dual channel, they can emerge in alternate sweep (one following the other) or in low-frequency chopped mode (500 kHz) so that both channels are synchronous during sweep. In the add mode, the two signals can be added and, with the inversion of one channel, channel 1 and channel 2 differences can be measured. In the differential mode, however, identical probes with no ground reference can monitor the voltage drop across floating components. However, a separate common to the oscilloscope from the circuit under investigation is required. In X-Y operation, accurate phase difference between signals of the same frequency can be verified up to 120 kHz, although the X-amplifier bandwidth is specified at 5 MHz. This can be proven by introducing an identical sine wave to both Y amplifiers and increasing its frequency until the diagonal (monotonically ascending) line begins to form an ellipse that indicates inequalities. A full circuit signifies a phase difference of 90°. Mathematically, the sine of phi (phase ϕ) is the ratio of the Y-axis crossings (a) to the total amplitude of the ellipse (b). The 90° angle, of

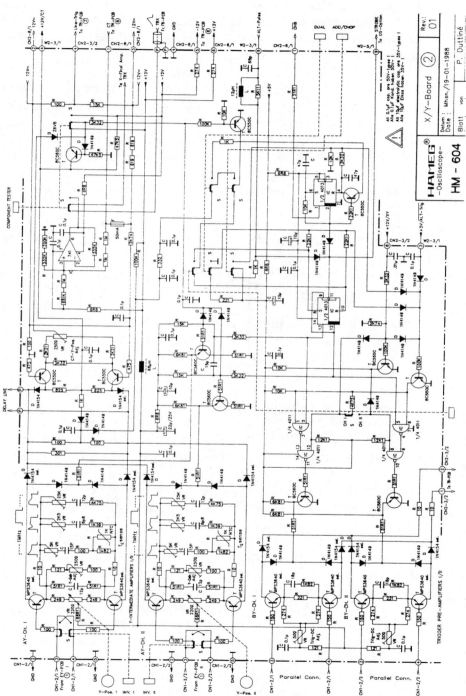

Fig. 3-5 *An actual schematic of the X-Y amplifier and logic board in the HM 604.*

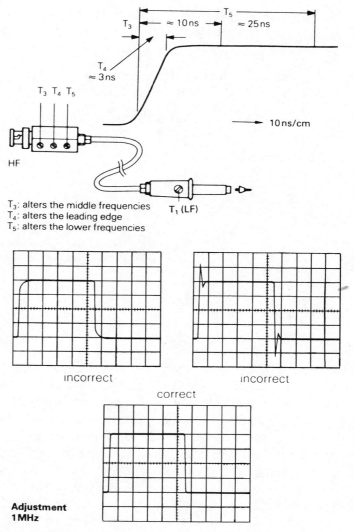

T_3: alters the middle frequencies
T_4: alters the leading edge
T_5: alters the lower frequencies

incorrect   incorrect

correct

**Adjustment
1 MHz**

Fig. 3-6 *Special probes often supplied with the HM 604 require careful high- and low-frequency adjustment.*

course, makes a and b unity equalities, and the arc sine of 1 equals 90°.

Phase-difference measurements are also possible at much higher frequencies using the two vertical Y1-Y2 amplifiers, when both the amplifiers and their probes are calibrated precisely. Alternate channel switching is used for frequencies above 1 kHz and attenuators are set for at least five divisions per channel. Here, a time difference of three divisions is selected against a total pair of alternations that extend the full 10 divisions of the horizontal X axis. Using t for the smaller portion and T for the larger, the following simple equation develops:

$$\phi = t/T \times 360° \text{ (degrees in any circle)},$$
$$\text{Therefore: } \phi = 3/10 \times 360 = 108°$$

The illustration in Fig. 3-7 justifies the calculation because 3:10 is the phase ratio.

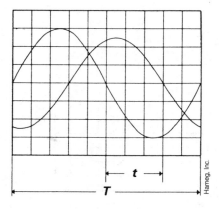

Fig. 3-7 *Accurate phase angles are easily measured with good equipment.*

Simple amplitude measurements are either peak-to-peak (p-p) or 0-peak, depending on the point of reference. As shown in Fig. 3-3, the sine wave's center is common (0) and you're only looking at peak. If the full measurement was required, the 15.6 mV would simply be doubled so that 15.6 mV becomes 31.2 mV. When working with sine waves, however, you must understand something more than just peak or peak-to-peak values, because other values are useful, too. Therefore, we photographed a full peak-to-peak value of a reasonable sine wave and called up a calculated readout of slightly more than 4 divisions at 10 millivolts (mV) per division. We will call the 42.3 mV ac readout E max for maximum rail-to-rail voltage (Fig. 3-8).

Initially you see one-plus cycle of a typical sine wave whose instantaneous value (e) would be calculated as:

$$e = E_{max} \times \text{sine of } \omega t$$

with ω (omega) standing for the quantity 2π (6.28) Xf in seconds at some frequency, representing angular velocity in radians/second.

The effective or rms value of this voltage would then be established by: $E_{effective} = E_{max}/2\sqrt{2} = E_{max}/2.828$ (over one full

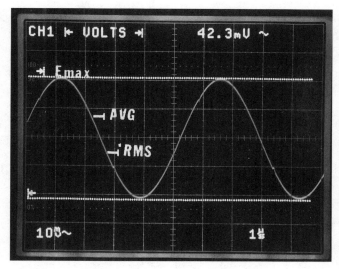

Fig. 3-8 *The three prime measurement conditions in sine waves include p-p (often shows as E^{max}, rms (0.707 of E_{max}, and average (0.637 of E_{max}).*

cycle), derived from the integral 0 to 2π. Instead of 42.3 E_{max} or p-p, the rms (effective) voltage would become 14.96 V rms, which is more than a considerable reduction.

You should also know the average value of a sine wave, which is actually the average value of a full-wave rectified sine wave over a full cycle (or half cycle): $E_{average}$ = 0.637 E_{max} × 42.3 mV, or 26.949 mV, and has nothing to do with the rms heating effect (as illustrated in the previous equation). It is, however, useful in dealing with a group of instantaneous sine waves that are separated by equal angles. Now, instead of working in simple p-p values, you have the measured concept of both average and rms sine-wave values, based on peak E_{max} voltage.

Pulses, though, are another matter completely and serve here as a suitable introduction to the subject of time and their relation to the horizontal (X) axis of the conventional oscilloscope. Such waveshapes are either a product of common-to-power supply switching logic or the formation of (usually) rectangular or square waves from a fundamental and its odd (3, 5, 7 etc.) harmonics—with the greater number of harmonics delivering the sharper and more desirable rise- and falltimes because sloppy waveshapes often produce false triggering or outright circuit failures.

We thought it might be constructive to superimpose a square

wave and a rectangular wave on the same printout (indicating their single-cycle cursor measurements), and then calculate the rest. Without cursors, of course, you'd have to estimate several parameters, but time/division and amplitudes are also recorded, so fairly simple math is in order (Fig. 3-9).

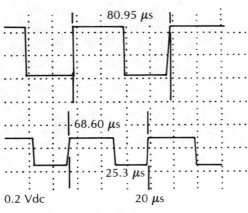

Acquisition mode: peak detect

Fig. 3-9 *Pulses and rectangular/square waves are easily characterized with a few simple cursor measurements.*

Symmetrical square waves always have a 50% duty cycle, so the frequency would simply be:

$F = 1/T = 1/80.95 \times 10^{-6}$ or $0.012\,35 \times 10^6 = 12.35$ kHz;

E_{max} at 0.4 V; and at 50 Ω Z, and power required would amount to:

$P = E^2/R$, or $0.16/50 = 0.003\,2$ W or 3.2 mW

Asymmetrical rectangular waveshapes might share the same general characteristics as their sister symmetrics, but they are considerably different. The lower, asymmetrical voltage has a 1-cycle time period of 68.60 μs, indicating a slightly higher frequency of 14.667 kHz, with a constant amplitude of 0.24 V, and a minor pulse width of 25.3 μs.

Here, however, we already know the measured times of both period and pulse, what's left is the waveform amplitude and necessary driving power. First, the average value (E_{av}) can be found by multiplying the pulse height by its duration (t) and the time

(T) of one cycle. It goes like this:

$$E_{av} = E_m t_d/T$$
$$= 0.24 \times 25.3 \times 10^{-6}$$
$$68.6 \times 10^{-6} = 0.088\,5$$
$$= 88.5 \text{ mV}$$

Similarly, the square root of the average value now becomes E_{rms}.

So, $E_{rms} = E_m\sqrt{t_d/T}$ which, simplified, becomes 145.75 mV. Notice that in this instance we're not dealing with a pure sine wave and a 2π full alternation. Therefore, E_{rms} is greater than E_{av}—almost the opposite of the numbers called out in Fig. 3-8:

$$P = E^2/R = 145.75 \text{ mV}^2/50 \text{ } \Omega = 0.425 \text{ W}$$

The frequency of a single cycle for the lower pulse signal is easily found by:

$$F = 1/T$$
$$= 1/68.6 \times 10^{-6}$$
$$= 14.58 \text{ kHz}$$

Work with a few self-generated examples using a few whole numbers, then milliseconds and microseconds, and the entire process should become reasonably clear. Or simply buy a 2252 Tektronix with printer and forget most of the calculations if a few thousand greenbacks are handy. Remember, however, that oscilloscopes don't produce amplitude readings in terms of rms; you must make a little effort for these.

After working through these useful, but somewhat tedious concepts, you should now be ready for the promised time-base explanation, which is somewhat more involved than simple voltage amplitudes.

The 'scope's important time base

Sine wave and rectangular displays are worthwhile, indeed, but without hard time-base numbers to back them up, such signals could have only half meaning and many of their important characteristics would never be known. Consequently, an oscilloscope that offers full utility must operate as an ac voltmeter, a dc voltmeter, a timing mechanism, and a two- or three-place frequency counter, depending on specifications. With a Hall-effect probe, a scope can also display current, then operate in the X-Y mode as a handy vectorscope. So, adding time to amplitude does wonders for convenience and efficiency.

The HM 604 time base is complex indeed and it represents much of our need in the foreseeable future. At an ac bandpass of 60 MHz (– 3 dB down), the instrument will trigger to 80 MHz and has an X-amplifier passband of 5 MHz—all worthwhile characteristics for low-cost display equipment. A drawing of this portion of the front panel appears in Fig. 3-10.

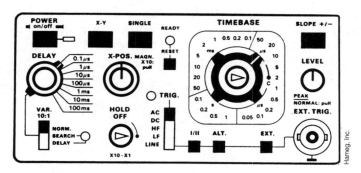

Fig. 3-10   *A front-panel illustration of Hameg's HM 604 illustrating time base, triggering, X-Y, and pseudo-delay.*

Although the scope can easily display almost any incoming signal up to 60 MHz, rectangular-shaped signals often have harmonic content that requires viewing also. Therefore, a 60/10 ratio, which is also a factor in digital storage scope megasamples—sometimes reduces the repetition rate to 6 MHz for reliable evaluation. Complex video signals usually need special sync separator triggering for line or field stability on either positive or negative slopes.

This active TV sync separator removes horizontal and vertical pulses from composite video so that weak or distorted television information can be seen. In the horizontal mode, trigger slopes are normally positive; but vertically, a low-pass integrating network is added to receive the six vertical pulses as well as their equalizers to make either field 1 or field 2 viewable. If video or sync is inverted, inputs remain undisturbed and only CRT plate signals are reversed. Ordinary standard video is always automatically triggered in the AT (automatic triggering) mode, but the normal mode has no display without signal input.

If a triggerpoint becomes difficult on these complex signals and level and normal triggering modes won't suffice, the hold-off circuit may be engaged to vary time between alternating sweeps in a ratio greater than 5:1 so that a stable sync point can be found.

Also in the dual channel alternating mode, there's an ALT button in the X-section that will alternately deliver trigger voltage from CH1 and CH2—a mode that is especially useful in synchronizing a pair of asynchronous inputs when in the normal condition only. Common sync prevails in the AT mode.

Additionally, depending on incoming frequencies, there are ac, dc, HF, LF, and Line (60 Hz) trigger couplings available for capacitor, direct, high-frequency, low-frequency, and ac power inputs (all filtered inputs with the exception of dc and line) so that the appropriate triggers are passed to the sync circuits. Also, a waveform ±slope potentiometer aids the triggerpoint either positively or negatively. When internal triggering won't do the job, an external trigger input is available to help. However, you should be able to trigger almost any signal with internally generated sync, because I have yet to use an external trigger with this scope.

Now, to view some tricky voltages in detail, the delay trigger switch should initially remain off, center the trace in its graticule and display between one to three cycles of the repeated signal. It is now possible to start the sweep at any desired portion of the waveform, expanded at least 100× (or more with the ×10 magnifier) for as long as you can view the trace. This viewing is done with the delayed switch between 0.1 μs to 100 ms and the norm-search-delay control and the 10:1 adjustable delay pot (a 20-turn precision potentiometer). Hold-off is set to minimum, with variable pot to cal, and level-adjusted for a jitter-free display. The mode switch is then moved to search and the display start shifts right, which denotes the exact time delay. If not, rotate the delay switch downward from 0.1 μs to a point just before the portion that you wish to be delayed. With the 10:1 variable control, select whatever is needed, and switch to the more permanent delay position. The search LED then stops flashing and remains fully on.

Just to illustrate the actions of this pseudo-delay, a triple exposure of a 3-field video waveform might clarify the procedure if demonstrated in other than strictly scope manual-writer talk. Figure 3-11 is actually a triple Polaroid camera exposure of the several steps taken. After turn-on, triggering is set to LF (low frequency), the TV Sep is adjusted to V±, and the Delay Trig is adjusted for Search. At that point, the precision potentiometer can select any portion of the waveform that requires inspection (as shown in the middle display). Thereafter, the ×10 expansion

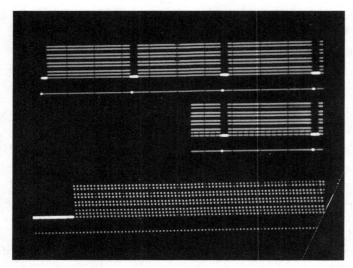

Fig. 3-11 *Pseudo-delay selecting one field, then expanding the third for detailed viewing.*

produces the selected area in full, aided by a little-added intensity and whatever fixed delay that's needed.

Although not precisely the same as the usual A delayed by B technique that is ordinarily attributed to dual time-base oscilloscopes, this pseudo-delay seems to operate just as well without requiring a considerable amount of additional and expensive circuitry, and certainly the video field (found so often difficult to sync by many, many competing scopes) is rock solid in all respects. If all this strikes you as a plug for the HM 604, that's exactly what it is! Hartmann, Merget, and Silverstein have built a very useful, cost-effective instrument.

The pseudo-delay actually replaces the second time base that is available on more expensive scopes and achieves virtually the same purpose—some portion of the trace is expanded so that it can be studied in detail. A look at the time-base modular PC board can offer some insight into the complexity of these various timing operations. Most semiconductors are discretes here, but one dual-purpose integrated circuit (a TL082) serves as a dual op-amp switch for the external trigger input and FSO. Operating voltages are apparently – 5 and – 12 V (see Fig. 3-12).

Most everyone knows the operation of the X-Y switch that engages the X-amplifier for comparing waveshapes with the Y amplifier, except that the X-amplifier, although having a bandwidth of 5 MHz, is only phase-matched to the Y-amplifier up to

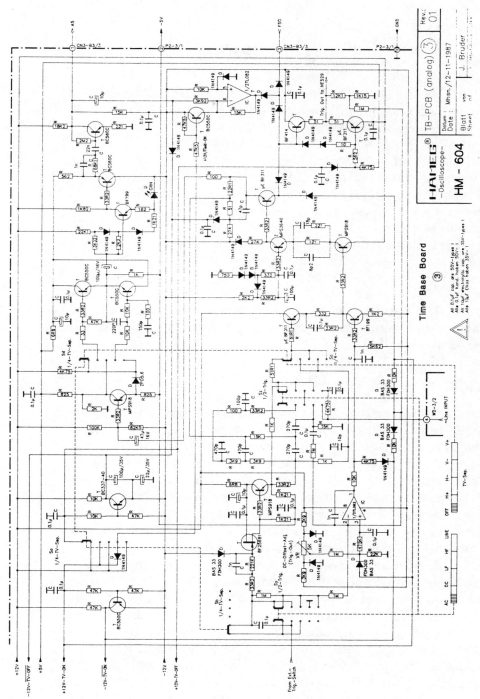

Fig. 3-12 *Time base and switching PC board under discussion.* Hameg, Inc.

about 120 kHz. Therefore, the X-Y phase function is limited to well below 5 MHz, and even a few hundred kHz. With both amplifiers capable of inversion, the unit is invaluable in low-frequency vector displays, which we will certainly illustrate in due time.

Finally, the HM 604 also possesses an electronic component tester, which is said to offer a go/no-go check for resistors, capacitors, inductors, diodes and transistors when removed from their circuits—with capacitors, fully discharged. Results, of course, are limited, and resistors are restricted in value from 20 Ω to 4.7 kΩ. Hameg offers waveform examples in its operating/repair manuals. Open circuits produce a single horizontal line; shorts develop a straight vertical line, with operating conditions somewhere in between.

That's about the story of ordinary oscilloscopes, including most of what's required in today's video, MATV, and cable businesses. The next step involves extraordinary oscilloscope(s) with printout, followed by digital-storage (DSOs) scopes in the next chapter. The foregoing should update most of the usual treatise on antiquated scopes, except for a few that are now mostly microprocessor-controlled.

Smart scopes w/printouts

Although a number of manufacturers have produced and marketed 4-input oscilloscopes at various prices and abilities, a good combination is difficult to find—especially if it's microprocessor-controlled, has programmable function LEDs, and has a matrix printout. We've therefore chosen to discuss Tektronix's 2252 (Fig. 3-13) as our model for the smart-scope portion of the chapter because of its remarkable flexibility and 12-bit A/D digital readout. By singling out one particular product, most or all its special features can be examined as examples are photographed and reproduced by the accompanying Epson printer on ordinary print paper. The combination, we believe, is rather remarkable. Such an instrument also offers a worthwhile opportunity not only to personally advance with the state-of-the-art, but also to learn new design and servicing techniques that are just becoming available. The 2252, therefore, is judged to represent the better product at this time and should model for a considerable number of copycats in the future, especially those just beginning to use microprocessors, menus, and lighted-mode functions. This

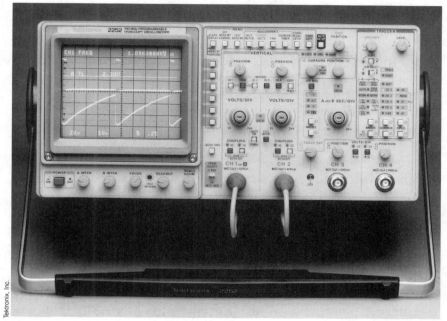

Fig. 3-13 *Photo of the 2252 programmable analog oscilloscope with digital printout.*

trend will obviously continue into the 1990s, because analog oscilloscopes with digital features have proven highly market-receptive already. For one thing, they're often less expensive than all-digital instruments and serve the general purpose, except they're not set up to collect and store one-shot occurrences for eventual external readout. For this operation, a DSO is mandatory.

The 2252 is described by Tektronix as a 100-MHz bandwidth, dual-sweep, four-channel portable oscilloscope that weighs 17.9 lbs. with microprocessor control that's menu-driven for voltage, frequency, and time measurements, and includes counter/timer and Smart Cursors™. It also contains an auto front-panel system setup that has both store and recall, menu and readout displays and internal calibration, plus a GPIB bus for controllable programming. The graticule is nicely lighted and its various functions are both plainly evident and LED illuminated. Once you use one of these advanced system oscilloscopes, it will be almost impossible to revert to standard equipment—especially if prices are even faintly competitive.

Basic deflection factors (without $10 \times$ probes) are 2 mV – 5 V/div. for the two main (1,2) channels and 0.1 – 0.5 V/div. for the other two, which are specifically designed for digital and trigger signals. Also, a $10 \times$ magnifier sweep switch can extend the fastest sweep to 2 nsec/div. With magnifier off, the A sweep extends from 0.5 sec/div. to 20 ns/div., and the B sweep begins at 5 ms and ends at 20 ns/div. Maximum amplitude for the two main channels is 40 V, or 400 V with $10 \times$ probes (Fig. 3-14). The auto setup triggers and the vertical/horizontal setup produces a waveform simply with the press of a button. Following this you can measure period, frequency, trace dimensions and rise- or falltimes by pushing another couple of buttons. Smart cursors then adjust to all horizontal or vertical measurements in question. Phase and propagation delay times are also possible, and it even has up to 20 front-panel setups in memory to help out for rapid and accurate responses, including standard service routines.

Finally, you can photograph the results on a lighted graticule or connect a hardcopy printer that responds to a 12-bit AD converter for a permanent photograph or printed readout. The overall vertical accuracy is specified at less than 3% and time base accuracy at under 0.8%, with an accelerating CRT voltage of 16.5 kV.

Such characteristics should be enough for accurate results and minimum capital outlay. Probes are included, as is an auto-level mode that automatically adjusts the 10% and 90% positions for automatic triggering while Mode buttons determine operations for the A and B trigger systems. The usual hold-off potentiometer sets delay between the start of A sweep, with respect to B sweep.

Triggering controls and functions are probably the most difficult aspects of the 2252 for anyone working with this instrument for the first time (Fig. 3-15). A trigger source is derived for either the A/B Select in the vert mode, which includes any of the four channel inputs or the 60-Hz line frequency, as selection indicators are moved up or down by the source buttons and trigger couplings that include DC (direct coupling), noise (reject), high-frequency (HF) reject (frequencies above 50 kHz are attenuated), low-frequency (LF) reject of frequencies below 100 kHz with dc blocked from the trigger, and ac for trigger-signal blocking below 50 Hz, which rejects any dc component.

All this is not too different from conventional scopes, except that four channels are to be triggered—in addition to the three

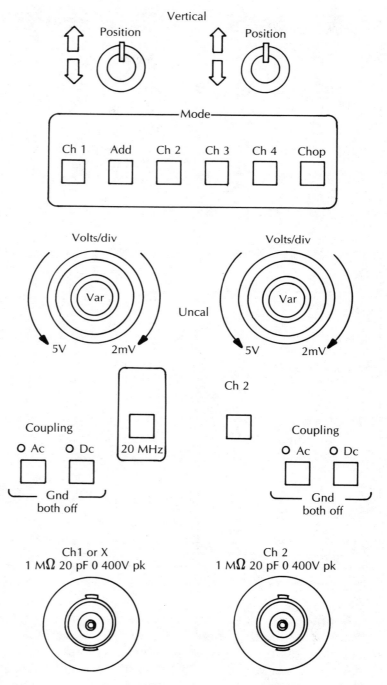

Fig. 3-14 *Front-panel drawing of the 2252 illustrating digital and analog functions. Notice the lighted control signals.* Tektronix, Inc.

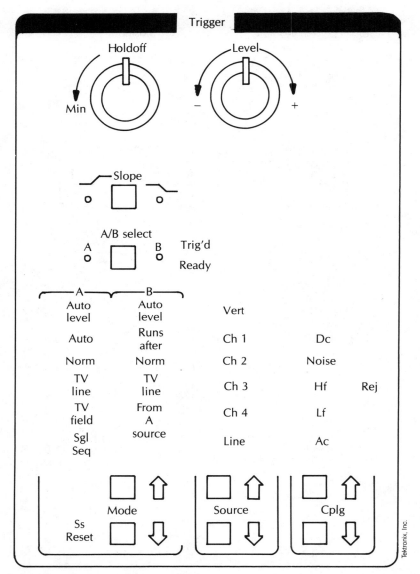

Fig. 3-15　*Trigger, slope, channel, and coupling selectors for the 2252.*

Noise/HF and LF reject conditions, which are phased somewhat differently by competing manufacturers. The same is generally true when the main A/B select is engaged, except that A and B means A and B sweep and not the individual time bases. When dual-channel C/T (counter/timer) measurements are undertaken, A/B select cycles through A, B sources 1 and B sources 2, when

A button is selected and lighted and B button is dark. Slope and level controls select ± signal slopes and the amplitude signal switch takes off.

The next subject involves A and B sweeps, trace separation, and the various modes, which involve operational selections for horizontal deflection(s). In Fig. 3-16, A and B sec/div identifies to two A and B time bases, with the A sweep speed set by the 0.5 sec to 20 nsec/div. detented switch, and the B speed always set higher than the A sweep. In the alt mode, sweep alternated between A and B is intensified, but only the B sweep can be adjusted with the A/B select set to B trigger. B sweep starts are delayed from A sweep trigger starts by setting the or delay time control, and B sweep traces can be positioned vertically, relative to A sweep when in the alt horizontal position.

At the top of the illustration are the cursor positioners for time, voltage, and frequency due for discussion shortly, as the horizontal position control. Below are the LOCK, SQR, and ADDR indicators that represent GPIB bus status. *Lock* means that the 2252 front panel controls are locked out and can be cleared only by "cycling" the power switch: SQR denotes a service request from the controller, an error or warning condition; the ADDR operates when GPIB is being addressed.

The various measurement modes are illustrated in Fig. 3-17. These prime the microprocessor for the various menus that display volts, time, amplitudes, and frequency, then print the results on hardcopy via an external printer. The measurement channel, for instance, can be custom-set, the display can be cleared or filled with commands, the voltmeter and counter/timer conditions can be posted, and the store/recall can place information in memory or extract same. Meanwhile, an auto setup button is available to automatically adjust front-panel controls for a usable display, and the hardcopy messages the printer to record the existing display on standard typing or photocopy paper. Suitable printers are the HC-200 (made for Tektronix) or the Epson FX series (either 9- or 24-pin series), available directly for Epson.

The FX-850 Epson

Epson America, Inc., Torrance, California, considerably loaned this deluxe FX-850 (Fig. 3-18) for the express purpose of evaluating printouts from Tek's 2252 portable oscilloscope. This unit has both continuous and single-sheet feeds, character fonts and

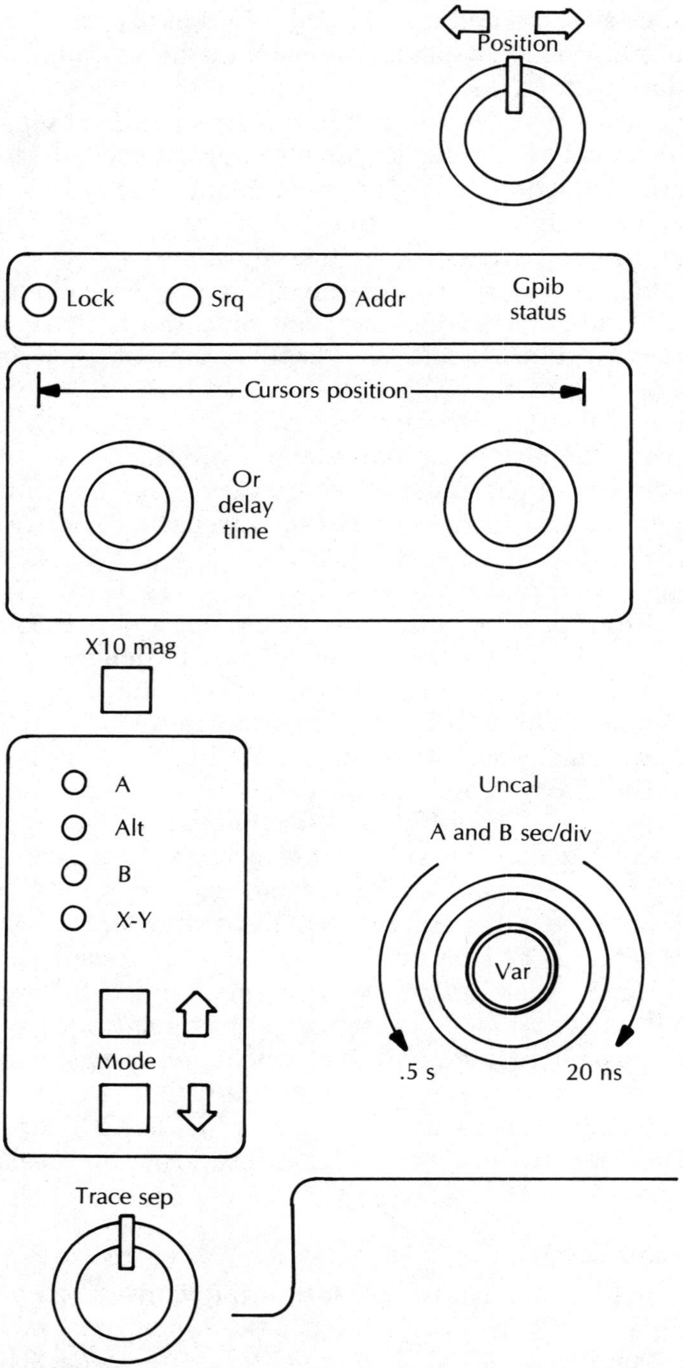

Fig. 3-16 *Mode, cursor, and time-base controls for the 2252.* Tektronix, Inc.

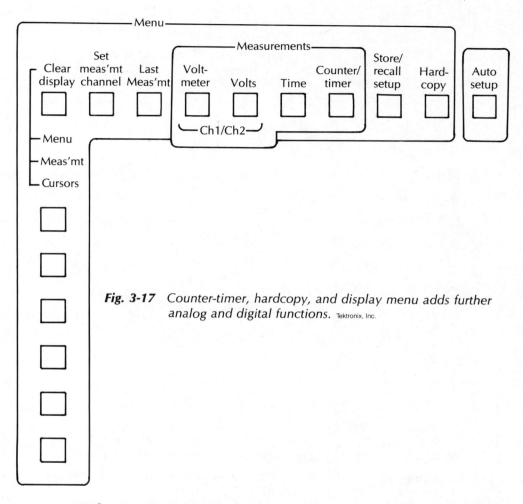

Fig. 3-17 *Counter-timer, hardcopy, and display menu adds further analog and digital functions.* Tektronix, Inc.

pitch, 190 characters/sec at 10 cpi, high and wide printing for headings and emphasis in addition to envelopes, an IBM emulation mode for IBM formats, plus the Centronics Parallel Epson-compatible interface that we're currently using. Other optional interfaces are also available.

The printer's setup booklet is quite complete, and once levers, shipping security, and cabling is complete, on-line and load/eject levers place the ordinary paper in the print position. Then (following cabling hookup), all that's necessary is to press the print button of the command unit, in this instance, the 2252. Other normally set controls are paper thickness, draft, roman, or sans-serif type at 10 or 12 cpi. Line and form feeds are lamp signals, as are ready and paper out. Dip switches behind the platen

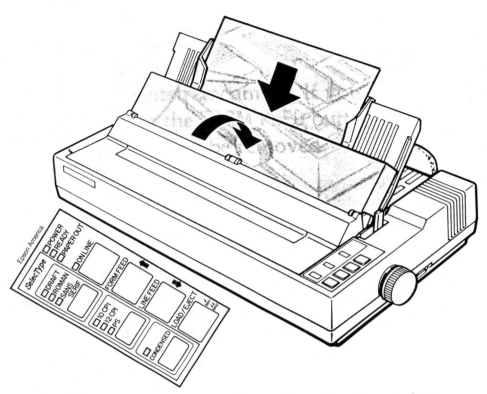

Fig. 3-18 *The FX-850 series printer, its control panel, and lamp indicators.*

knob will also affect character set and page lengths, along with a table of international character sets and other convenient and special adjustments. In the condensed mode, characters are printed at about 60% of normal character widths. The FX-850 will operate with word processors, spreadsheets, produce detailed graphic images, accommodate computer ASCII codes, deliver dot graphics, and even operate in BASIC.

The printhead on this machine has nine pins and prints via a ribbon. Electrical pulses operate these pins as the printhead moves across the paper and its metal carriage or transport. As this occurs, pins produce small dots via the inked ribbon and letters, numbers, and symbols are imprinted on the paper. Because there are 256 possible combinations of the pins, each pin must be assigned a number, and by adding the necessary label numbers, any combination of pins will fire—from 0 to 255. Ribbon life expectancy is 3 million characters with 14 dots/character, and machine meantime between failures for the FX-850 is 4 000 power-on hours. User suggestions are contained in Epson's user application notes for dot-matrix printers. Needless to say, this equipment is very well-designed, accurate, and sturdy.

Applications

With all descriptions and operational theory reasonably complete on these three outstanding pieces of equipment, it's time to apply their many features and uses to everyday problems and solutions for your own guidance and individual applications. Ours are basically examples with accompanying descriptions, but yours involve specific concerns devoted to analog and digital measurements that might or might not be accurate. If, therefore, your instrument is properly calibrated, possesses adequate probes, and remains fully grounded throughout the measurement cycle, results should be satisfactory. With poorly calibrated instruments and nonrecommended probes, voltage and time displays are usually misleading, to say the least. Among low dc supply voltage equipment and those with noncomplex logic and the usual impedances, your problems are relatively few. However, equipment with high-speed interactive logic and higher voltages, easily induces spurious pickup—even from isolated circuits and adjacent subsystems. Where high voltage and switching supplies are, this same pickup problem—especially in television receivers and monitors, is almost always prevalent. So, be on guard for spurious voltages where levels are either excessive or unusual.

You must also be on the lookout for circuit loading from your oscilloscope—especially when using only a 10-kΩ "direct" probe. If, for instance, another 10-kΩ impedance node or branch is being measured, you automatically halve the signal amplitude and produce an immediately false readout. Again, no ground or an imperfect ground will often produce undesirable circulating currents or local signal pickups that interfere with a true display. Any circuit with a dc component will move the scope trace up or down, depending on its value. When you need to measure ac only, switch in the scope's dc blocking capacitor and there's no problem. However, especially when analyzing low-frequency rectangular voltages, use only dc, or rise- and falltimes will be seriously affected and cause erroneous waveshapes from low-frequency reactance. The same advice holds for probe calibrations. Use only the direct instrument input for accurate amplitudes and timing. You are also advised to make measurements with 3-ft.

probe leads rather than with those that are longer. This length minimizes capacitance and inductance, and produces a more accurate readout.

All the above suggests an oscilloscope reminder: scopes can measure ac, dc, time, the inverse of time (frequency), phase differences, and even AM modulation. Further, if you have an adequate source, scopes are also capable of displaying vectors as well as current, but in terms of voltage per mA and amps. Unfortunately, these scope-current probes, whether ac or dc, are expensive and a good digital meter will work as well or better if it has a four-place readout.

Just about all oscilloscopes today have eight vertical divisions that are crosshatched with 10 horizontal divisions; the first for amplitude and the second for time. Whether units of measurement or true centimeters/div., readouts can all be considered the same for general purposes, because theirs is a per-division arrangement, rather than one that is limited to metrics. The better scopes, indeed, are metrics, at least across their graticules. Adequate metal shielding for both the cathode ray tube and the scope's chassis has always been a necessity to keep outgoing radiation and incoming interference away from the desired signal. Traditionally, oscilloscope cathode ray tubes have measured approximately 5 inches diagonally. Nonetheless, some are now 6 inches or even larger in some instances. However, medium-retention phosphors and 5-inch CRTs serve most purposes well and amplifiers are readily available to handle both vertical and horizontal deflections, as selected by the operator.

CRT-accelerating voltages are another matter entirely; the faster the time base, the more second anode voltage is needed. For 100-MHz scopes, 16 kilovolts (kV) isn't necessarily overkill. For 60-MHz scopes, 12 kV will pass muster, but 14 kV would be just a little better. Trace photography is very difficult without an adequately lighted graticule. Many scopes (even some of the better ones) are inadequate for satisfactory waveform presentation along with distinct graticule crosshatch markings. Trace-spot size means the difference between sharp and fuzzy images.

You must consider many things when evaluating an oscilloscope. You want reasonable vertical deflection, an extended time base into the nanoseconds, good H/V accuracy, vertical amplifier inversion, and especially excellent triggering on composite modulation. Particularly, you need good power supply regulation; no image on the CRT should budge when the power mains are variac'd between 90 and 130 Vac.

At the moment, the foregoing winds up the advice and consent portion so that we can begin directly with the various applications that begins with the Hameg HM 604.

HM 604 examples

These examples will consist entirely of photographs because the 604 has no printout facility, although the video plug on the back could be connected to any TV monitor for enlarged viewing, but without benefit of a calibrated graticule. Therefore, all photos are frontface on the CRT where amplitudes and time base products can be seen and evaluated. Because this is a servicing scope, we'll begin with over-the-air baseband composites to give you a "feel" for actual images and correct measurements that you should encounter whenever working with real-time audio and video. The RF electromagnetic signals are discussed later in the spectrum analyzer chapter because ordinary oscilloscopes have neither the range nor a video detector to make such detected information reliably useful. Half-wave detector probes, as stated, are neither exactly precise nor reliable. Therefore, we proceed directly to video/audio demodulated circuits for our information and examples.

The initial illustration in Fig. 3-19 shows incoming RF from channel 7 at 20 mV/div., directly off the TV antenna, and the lower baseband TV-detected information in the vertical-blanking interval of 1.33 msec duration at 1 V/div. The slightly compressed intelligence to the right of the lower trace, the conventional vertical interval test signal (VITS) and the vertical interval reference (color) signal (VIRS), are no longer required by the Federal Communications Commission, but are still used throughout the commercial broadcast industry. They appear, as formerly assigned, on lines 17 and 18 for VITS and line 19 VIRS. In addition to this display, which will be expanded in the next illustration, you also see equalizing, vertical, and horizontal sync pulses, which occur between lines 1 and 9 of the vertical blanking interval (VBI). Try this on your standard conventional oscilloscope and shudder at the result.

In Fig. 3-20, VITS and VIRS are fully exposed—simply by slightly repositioning the horizontal scope trace and using the 10 × magnifier. The entire display is extremely helpful in evalu-

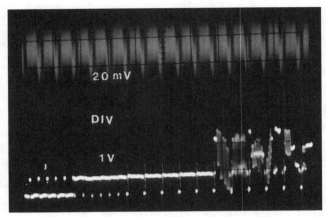

Fig. 3-19 *RF and baseband, sync and blanking intervals off the air.*

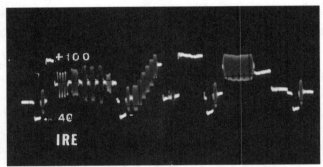

Fig. 3-20 *VITS and VIRS composite waveforms appear on scan lines 17 through 19 in the vertical blanking interval (VBI).*

ating video and other linear amplification from 5 MHz down when excited by such types of input signals. The – 40 to + 100 are the IRE units, which your oscilloscope processes at 1 V peak-to-peak. Beginning at the left, you see a horizontal sync pulse that is reasonably squared off, followed by a chroma burst, then six cycles of multiburst from 0.5 – 4.2 MHz (the latter down approximately 6 dB), another horizontal sync pulse (not too square, which indicates low-frequency roundoff), then a second color burst, followed by a modulated staircase that appears very linear and will produce excellent grayscale, the one 2T and one 12.5T pulse (not visible) are unmodulated and modulated (respectively) for B/W and chroma response characteristics and delays, preceding an 18-μs "window" that responds to any ringing or low-frequency rolloff (which there is). After the "window'' is another horizontal sync pulse, burst, and the + 40 IRE chroma reference signal, black reference, the fourth sync pulse, and more burst. All told, you're looking at three horizontal test pattern lines in the VBI interval that warns of low-frequency roll-off and a fair 4.2-MHz video bandwidth rolloff. Nonetheless, this is an awfully good reproduction of a broadcast set of test patterns—as seen at the video output of a Zenith SG2799 advanced System 3 monitor/receiver. Try this procedure on a $300 TV set and see what you get! Even the scope will blush!

One prime reason for test pattern abnormalities is as a result of the lack of a chroma/luminance separating comb filter. Any receiver that doesn't have one has a maximum video passband of 3 MHz and many can only offer 2.5 MHz. Therefore, multiburst is no longer visible at or before the fourth period of oscillation and both the resolution and the definition of the CRT image suffer. Projection receivers, with all their other problems, are especially sensitive to comb-filter omission or poor design, and cannot reproduce satisfactory pictures without them.

We might also add that these same or very similar test patterns are often available from geosynchronous satellite transponders in the video portion of NTSC fields and frames, are easily synchronized without special triggering, and could actually be used from satellite receiver video outputs to check TV/monitor receivers at baseband, and with lesser effectiveness through modulators via TV channels 3 or 4. This procedure is also very adequate for evaluating television sets by distributors and retailers before selling them to the public—a sure way to separate quality from price and quantity. Comb filters, whether glass delay lines with a few transistors or IC charge-couple devices, are very effective chroma/luma separators.

For occasional phenomena (such as one-shots), a camera with an umbilical attachment to the scope and 3 000-speed Polaroid film might solve the problem, but a DSO digital-storage scope would be the positive solution. This aspect is discussed and illustrated in the next chapter. A dot-matrix printer here, however efficient, won't unravel a fast one-shot for anyone, as long as it is simply connected to an analog scope.

On the other hand, digital displays with related frequencies are natural for dual- or quadruple-trace scopes and can be sync-locked and photographed with ease. In addition, some digitals that have asynchronous information can also be synchronized, but this is not a guarantee of most analog scope equipments and should be evaluated before becoming a firm believer. Even a pure logic analyzer could have some problems in this situation.

Vectors produced by 10-bar gated rainbow generators with basic oscillator timing of 3.56 MHz are also useful in TV receiver or monitor/TV evaluations, with signals injected through RF/tuner inputs. Such signals determine chroma demodulator and amplifier characteristics and can be most revealing. However, instead of appearing at video outputs, 10× dual-trace probes, which are connected to the red and blue cathodes of the picture tube, tell the true story. While you're working internally in the TV receiver, you might also want to pick up a multiburst at the video detector and also at the CRT's red gun. These will also confirm receiver passbands fairly accurately, especially if a first-class NTSC bar and pattern generator originates the signals.

A gated rainbow vector is illustrated in Fig. 3-21. This one, however, originates from a switch-modified scope that inverts both channels, rather than simply running single-channel inversion. In the latter instance, the pattern remains 90° out of phase, but it is still useful if you mentally complete the 180° rotation.

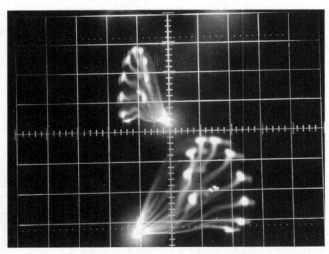

Fig. 3-21   *A gated rainbow vector imperfectly adjusted at 90° (top), and correctly at 180° (bottom). This test signal is highly useful, but it should adjust somewhat better than this.*

Formerly, before extensive use of much more precise integrated circuits, a fairly precise format was used for these vectors. Bar 9 was nulled at the red CRT gun (shown as R-Y) and nulled at bars 3 and 9 on blue (B-Y). Because the advent of grounded-grid

CRTs with luminance and chroma combined at the RGB cathodes, plus the addition of precise IC detectors, the "hounds tooth" representations at the bottom of Fig. 3-22 are considerably more even and virtually no nulling occurs. So to set up your vector display today, simply adjust your 10 × probes to at least 2 (20) V/div., connecting the scope's no. 1 probe to the red CRT cathode and No. 2 probe to the blue cathode. (Green is not used because of the constructed phase difference.) Then push in the scope's X-Y amplifier combining button and adjust the TV controls for a display as close to the lower trace in Fig. 3-21.

If you're still 90° out-of-phase (as in the top trace of Fig. 3-20), even with channel inversion, do the best you can with black level, picture, and controls to approximate the bottom vector (as illustrated). In it you will observe no petal crossovers; the hammer-head brighter illuminations develop as each arm of the vector traces its up and down swings, with reversals at the brightened portions of each of the 10 petals. If all are separated by 30° with general regularity, and no crossovers or extra distortions are obvious, then your color circuits are doing just fine. However, gross linearity problems, a compressed red or blue side of the vector, or other irregularities signal either bum chroma detectors or bad luma/chroma amplifiers, and that's that!

Low frequency audio, including the 100-kHz multichannel (BTSC-dbx) spread, is a cinch for almost any oscilloscope that has even reasonable measurement capability. Remember that analog scope accuracy is seldom better than 2%. Therefore, an accurate ac meter with the necessary frequency range might operate with somewhat better resolution. However, such a meter could cost a fair amount, and a low-frequency scope with many more uses certainly will offer more options for less dollars.

Sine waves and square waves should be rather familiar and even phase measurements are rather simple, but ac/dc combinations and time inversions might not be second nature to those who do not ordinarily deal with such seemingly trivial things. You should know this information because dc levels are often triggerpoints and ac amplitudes determine the extent of information that's processed.

A circuit that triggers on positive excursions, for instance, wouldn't do very well if its level is negative and below turn-on thresholds. Similarly, too much positive dc could saturate some ensuing analog circuit and kill the intelligence flow. Then again, a small ac signal that rides on a large dc voltage could be missed

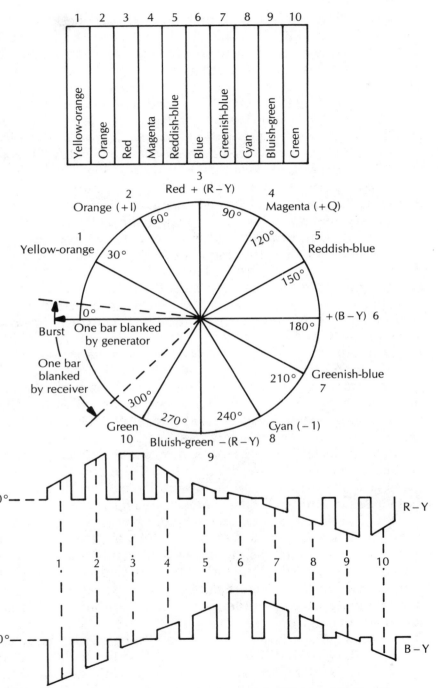

Fig. 3-22 *The gated rainbow vector display checks luma/chroma input to cathode ray tube.*

altogether, which is why many scopes (such as the HM 604) have off-scale warning LEDs and a trace-locator circuit.

You're also probably aware that rise- and falltimes of voltage steps (usually rectangular waveshapes) are products of the time difference between the 10 – 90% points and the time-base setting. OK for "slower" displays, very fast rise-/falltimes must take into account scope amplifier risetimes, as well as those of the probe. A typical example, as given by Hameg, amounts to:

$$T_{actual} = \sqrt{Tr_{measured}^2 - T_{amp}^2 - T_{probe}^2}$$

Where: Tr is the measured risetime, T_{amp} is approximately 5.8 nsec, and the probe risetime usually is 2 nsec. Square each identity, do the arithmetic, then take the square root of the result. Hameg says that if Tr is greater than 42 nsec, forget the calculation.

Now, work out a few samples for both frequency measurements and risetimes just for familiarity. Working with the leading edge of the "rectangular" waveform in Fig. 3-23 and the fourth vertical division from the left, observe exactly 3 cycles (leading edge to leading edge), which are multiplied by the time-base setting of 2 μs/div. The frequency, therefore is $F = 1/T = 3/1 \times 2 \times 10^{-6}$ or $1.5 \times 10^6 = 1.5$ MHz. Taken to four places, this measurement would count out to 1.556 MHz, rather than 1.5 MHz. So

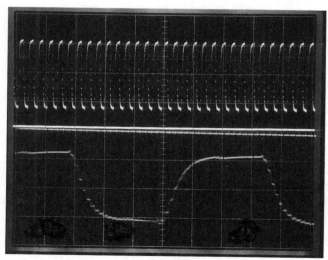

Fig. 3-23   *Double exposure illustrates frequency (top) and risetimes (bottom) of a sloppy rectangular waveshape.*

you can find an answer, but with no more than a 2-place scope accuracy.

Stretch this "rectangular" waveshape out to 0.1 μs/div. and it appears as a very sloppy high-frequency rolloff with risetimes and falltimes of approximately 0.1 μs. Calculate the display:

$$T\,actual = \sqrt{100 \times 10^{-9^2}\ 5.8 \times 10^{-9^2} - 2 \times 10^{-9^2}}$$
$$= 99.81 \times 10^{-9}$$
$$= 0.099\,81 \times 10^{-6}$$

This figure is close to 0.1 μs.

Return to the 1.5-MHz upper trace for a moment and notice that this waveshape is well above the "dc" reference below. How well do you think this would integrate that far above common? Put a capacitor in series with the output and you'd do well, but it would have no direct-coupled integration here with the waveform this much above dc.

All right, those are just a few of a worthwhile analog scope uses; many more are worth exploring. Meanwhile, it's time to proceed a step up in analog scopes, which are equipped with digital readout.

Analog scopes can have digital readouts

The Tektronix model 2252 appears just as a better-than-average 100-MHz analog, 4-channel oscilloscope, until you look closely. Then, you discover LED mode indicators all over the front panel, in addition to a 12-bit A/D converter (under the metal case) that converts this instrument into an inexpensive digital readout— good for almost everything that true digital-storage scopes are famous for. In short, for relatively modest dollars, you can have an extended-range analog instrument with high accuracy, microprocessor-controlled menus, and a very inexpensive dot-matrix printer that receives either average or peak-detect outputs that will pick up a remarkable number of anomolies and glitches, especially in both sine-wave applications and in digital logic.

Earlier, many of the 2252's characteristics were recorded as well as specifications. However, those only show what an instrument can do and not how it actually operates. Therefore, we thought a block diagram description might help to form an understanding of what's involved. Following such a complex block might be slightly confusing; microprocessors, multifunctions, related component interactions, operator commands, and large-scale ICs don't contribute to bed time stories, but we'll try to translate as simply as possible.

The Tektronix "simplified block diagram" (Fig. 3-24) should tell most of the instrument's story. It's divided up into 17 black-outlined blocks with signal-connecting lines in between as well as several large bus channels, so the scope can internally transmit and receive the necessary orders and data.

Block 1 is the four-channel input for signal flow and triggering. Although channels 1 and 2 respond ac or dc to signals within 2 mV to 5 V/div., channels 3 and 4 are limited to 0.1 V and 0.5 V per div. and are dc coupled only. Although channels 1 and 2 have variable attenuator settings, channels 3 and 4 are buffered by high-impedance FET amplifiers and their gains are fixed by the preamp adjustment only. Most signals go to the vertical preamps and output amplifier (2), but the control mux (7) and the display and trigger logic interface (4) supply the inputs. Outputs from the preamps and amplifier go to the A/B trigger system (3), the display and trigger logic and processor interface (4), the A/B trigger system (3) as well as ±plate-driver signals to the cathode ray tube.

The A/B trigger logic interface possesses a number of inputs (and delivers outputs), such as trigger levels, bandwidth limiting, calibration, T^2L clock, delay, trigger clock, line trigger, and zero HYST. The A/B trigger logic/interface delivers A and B triggers, vertical compensation and X-axis magnification. When trigger levels appear above their comparator's threshold (set by the level and slope controls), the comparator output changes state to start the A/B sweep. Then, the B-trigger becomes a successive approximation D/A converter to signal peaks or dc levels. For video, the trigger detector selects either a peak positive or negative, which sets reference input voltages that are sufficient to strip video and expose composite sync. The measurement processor (8) determines the ensuing trigger modes and the control bits do the rest. Affected are A/B triggers, TV line/field, auxiliary data shift, etc.

The various operating modes of trigger circuits are controlled by the measurement processor (8), which operates with control bits that are loaded into shift registers for the switch board (10), the DAC (digital-analog converter), display and trigger logic (4) the A/D converter (11), and the readout system (9). The logic processor (4) interface operates serial communications

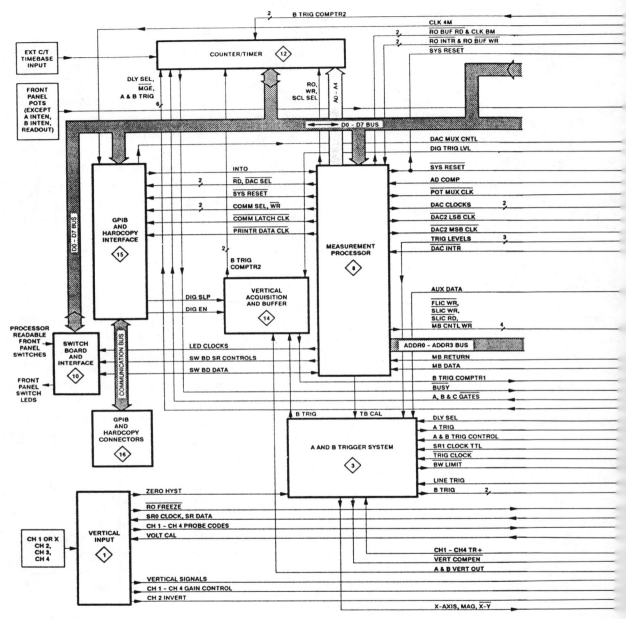

Fig. 3-24 *Signal flow block diagram of the 2252, which shows the various block functions.* Tektronix, Inc.

between serial shift registers and the measurement processor (8). Control data loads on the ADDR0-ADDR2 bus lines latched, in turn, on the SR DATA signal line. A and B sweeps (5) time and produce A/B ramps to drive CRT deflection plates. Hardware

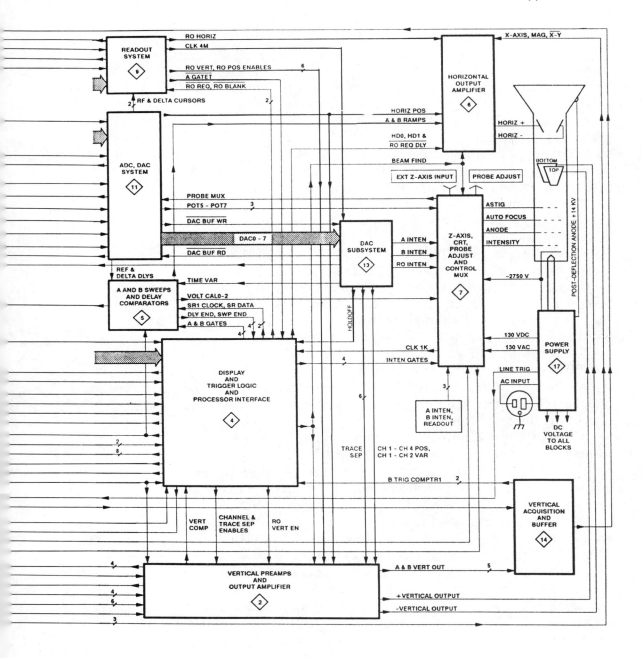

states derive from control bits that are loaded into controlling shift registers. Separate registers store A-sweep and B-sweep timing bits. Driving signals first pass into the readout system (9) which, in turn, controls text display and cursor readouts, as determined by the measurement processor. ASCII character codes in the RAM (random-access memory), through the dot-positioner ROM, are then converted to D/A signals via the ADAC (11), on to the horizontal output amplifier (6), and to the cathode ray tube.

In addition to V/H plate drivers, a Z-axis input is from block 7, where level-shifting occurs to match the negative cathode voltage of the cathode ray tube, which is aided by auto-focusing and an intensity control. A probe-adjust signal for the scope's front-panel probe calibration is also generated, and a scale illumination voltage forms front-panel drive for graticule illumination.

Under control of the measurement processor, the DAC subsystem (13) converts digitized control voltages to analog for the vertical preamps and the display, and trigger logic and processor interface, in addition to the A-axis IC already identified as block 7. The vertical acquisition and buffer (14) supplies B-trigger for the counter/timer in block 12. This unit contains a 10-MHz crystal-controlled oscillator, an external time base-input, two counters, level translators and a phase-locked loop. Here, period, width, and delta time modes are generated.

Basically, that's how the 2252 operates—with the exception of GPIB hardcopy interface and connectors 15 and 16, respectively. These are the 2252-Epson printer interfaces, which allow paper printouts of selected waveforms via the external port on the case of the instrument.

Granted, this operational description could have been more complete, but it might have generated undesirable confusion and have required insertion of certain schematics and other material that would be rigidly technical and possibly of interest to only a very few. The power supply (17), however, evolves from a switch-mode circuit that functions from 90 to 250 volts and from 48 to 445 Hz although the cathode ray tube receives a + 14 kV acceleration voltage (with reference to a – 2.7 kV supply to the cathode), which delivers an overall second anode acceleration of 16.7 kV. That circuit and system talk should be enough to satisfy most. The remainder of the chapter consists only of system printout examples and short explanations of each.

Pictures are worth a thousand words. Therefore, let's generate a few and see if that old saw tallies with oscilloscopes.

A sawtooth isn't always linear; the one in Fig. 3-25 positively isn't! Somewhat ragged, according to the printout, the trace appeared very smooth on the scope trace with none of the aberrations apparent in the slightly jittery down and up-slopes.

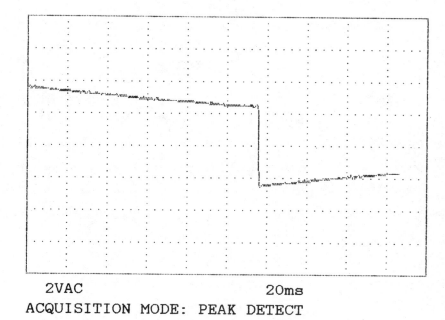

2VAC   20ms
ACQUISITION MODE: PEAK DETECT

Fig. 3-25   *A somewhat suspect sawtooth voltage with visible jitter with TV field stabilization.*

However, we did have to go to the TV field setting for stable sync. Obviously, the scope produced the display and hold, while the printer took its memory cue and did a reasonable job of reproducing a vertical driver waveshape. The points to cover numbered 180.

Now speed up a cruddy sine wave where your oscillator can supply plenty of spurs or even ragged modulation, and the printout appears (Fig. 3-26). It isn't a pretty sight, but in the peak detect mode it shows that the printer might be seeing what the scope's CRT doesn't reveal. And at 200 ns/div., the value of a single cycle measures 0.579 5 μs, which proves that the signal generator was virtually out of range and so much garbage on the trace might well be expected.

Look at the dual trace layout with 468 points to acquire (Fig. 3-27). The more points there are, the longer it takes the printout logic to follow trace outlines and put them on paper. In this case, you have two waveforms; that means that 936 points are in sequence to pick up and store before printout, so you'll have a little wait before your copy paper tells all.

The first trace in Fig. 3-27 doesn't reveal much, not even that the overshoot portion of each waveform was slightly rounded,

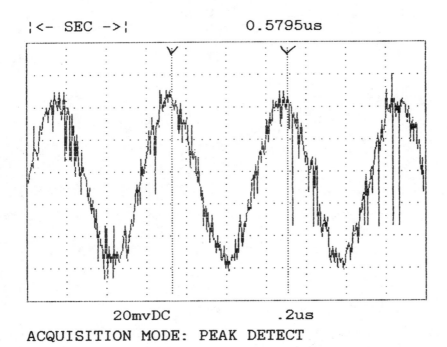

¦<- SEC ->¦   0.5795us

20mvDC   .2us

ACQUISITION MODE: PEAK DETECT

Fig. 3-26 *A dirty sine wave, alternating along at 1.7256 MHz, has many*
faults, plus noise and glitches. The markers are arrow'd.

indicating some high-frequency loss or a poorly adjusted
probe—probably the latter because our generator had good ICs
with no capacitance involved. The second staircase voltage, however, appears linear on the scope trace, but is full of very sharp
risetime and falltime ''icicle'' glitches between the formations of
several steps. Observing this disparity, we went back to the scope
and turned up intensity for maximum brightness. Bingo! These
anomolies did indeed appear; all of which suggests that high
intensity needs to be used on all questionable voltages if you are
suspicious that problems such as these exist. Better yet, hook up
your printer and let it be the judge. Apparently, this advantage is
significant over the eyeball and tube; it might be your salvation
when some circuit is false triggering and the answer isn't apparent. In the average (nonpeak detect) mode, such glitches would
not be obvious. Therefore, use peak detect throughout this portion of the chapter, just to be safe.

However, don't expect these printouts to reveal everything. In
complex waveshapes, they aren't reliable (at least the low cost
ones) and the faster the frequency, the lower the reliability.

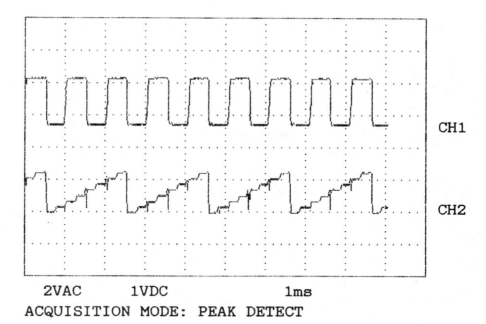

2VAC   1VDC   1ms

ACQUISITION MODE: PEAK DETECT

Fig. 3-27 *Here, 468 points for each trace at 1 ms/div takes a long time to acquire and print. Note glitches in chapter 2, but chapter 1 did not show high frequency rolloff.*

Therefore, when the trace to be copied and printed first appears on the face of the oscilloscope, observe it carefully, then push the print button with a pretty good idea of what you expect to see. In this life, very few things or people are perfect, and if it was not for imperfections, many of us would be out of business. So, welcome a glitch or two into the family and keep searching for its origin. Some you might never discover!

This series could be continued ad infinitum, but by now you should be sufficiently introduced to the art and design of dot-matrix printing. Some oscilloscopes, especially the DSO group investigated next, have more than one format available, and it might be that X-Y copiers backed by plenty of storage can render improved service, compared to the dotters. Again, ask for a good demonstration before you buy!

❖ 4

The bits and bytes of digital storage

ANY ARTIST HAS DIFFICULTY WITH MOVING SCENES, BUT HE OR she is usually content with immobility. Engineers and technicians often face the same problem, especially when analog triggering results in jitter, partial information, and sync loss. Enter, therefore, the digital-storage oscilloscope, which can look at single-channel or multichannel information, stop it in its tracks, and allow many definitive measurements to occur, up to and including the parameters of the instrument.

Commonly known as DSOs, these storage scopes today are often part analog and part storage, but some manufacturers have ceased making analog scopes entirely and are counting on new designs and freeze-frame (stored) ability to carry the load as megasample rates become even more rapid than they have been in the immediate past. Previously, megasamples were something of a miracle, but already top products are a factor of 10^3 better, as faster logic and A/D/A conversions creep up the multibit ladder. Certainly, they won't stop at mere megahertz. As time and costs are amortized, digital-storage instruments will certainly supplant their analog kith and kin in most (if not all) endeavors, except the least-expensive student-type applications.

Coupled with inexpensive dot-matrix or better-resolution X-Y plotters, the price of a printed readout will be pennies, rather than dollar(s), per copy for film. However, if you want maximum detail, rapid sweep information, or stark contrasts, film still prevails as the preferred medium—especially if you are fortunate enough to possess one of the new Polaroid DS-34 cameras with the appropriate, matching hoods. The definition and resolution in these recent, all-mechanical cameras is remarkable—as you can see throughout the book, and especially in chapters 4 and 5. For printers, be aware of special connectors, though, because specific logic matches are required for equipment, such as Epson or IBM formats that the DSOs can accommodate. However, good 9-pin dot-matrix printers in many formats are now available and are fairly inexpensive, so all you have to do is ask. Now, here's the rest of the DSO story.

Keep it simple

Once upon a time, DSOs were mysterious, expensive, select oscilloscopes that were equipped with slow-speed magic and all sorts of knobs and buttons to confuse and challenge their users. Today, the distinct trend is toward a basic extension of the analog domain with sufficient storage and sampling ranges to attract both those engaged in analog, as well as digital investigations. At the moment, many such instruments are half analog and half digital, allowing reasonable measurements to 100 MHz or more analog and 10 MHz of passband digitally. As A/D converters improve, this rather poor digital ratio will progress (it already has in some expensive instruments) until much higher bandwidths can be sampled and displayed. Meanwhile, with a flood of DSOs due continuously on the market, pick and choose carefully. Some are considerably more desirable than others, and the really low megasample units are hardly advantageous at all. DSO leaders in the mid-price ranges, such as Hewlett-Packard, Tektronix, Philips/Fluke, and some of the Japanese, still maintain a considerable share of the market, based on the usual criteria of price and performance (Fig. 4-1).

Once demand has made manufacturing relatively profitable, you can expect competitive price decreases, enhanced features, and considerably more acceptability across the electronic spectrum. It's strongly indicated that a full-featured spectrum analyzer, a well-designed DSO with two or more channel inputs,

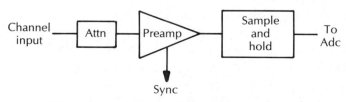

Fig. 4-1 *Digital scope sampling procedures.*

and possibly a multifunction generator can produce considerable insight into both analog and digital systems and subsystems. Just make sure that all drive voltages, CRT displays and sync requirements are fully satisfied. As you will shortly see, DSOs with cursor and time/frequency, amplitude, and rock-steady stored readouts beat a simple ac/dc meter every time. Fortunately or otherwise, we will all have to grow with the technology, because computer-aided work stations (CAWS) are actually already on the way.

Historically speaking

Just a scant 10 years ago, digital storage oscilloscopes first began to be somewhat noticed. It was then that such terms as: *roll-mode screen display updating*; *pretrigger storage* which defined the amount of information accumulated in memory before triggering occurs: *memory expansions*, which usually means a 10 × vertical/horizontal elongation; *stored writing speeds* for analog-storage scopes; and *equivalent time sampling* to permit repetitive waveshapes to be sampled during multiple sweep scans, etc. At the time, manufacturers were just working through the stepping era of bistable scopes, where CRT phosphors consisted of the prime storage and approaching all-digital instruments were little more than expensive hobbies. The pattern, obviously changing for both low and high frequencies, has brought about a continuing radical departure for both analog and the none-too-successful CRT-storage equipment, where storage did, indeed begin, but analog, foggy readouts and fast-button pushing soon made digital on-screen displays somewhat practical. This practicality spurred further development, which often combined analog and digital instruments, until recently. Now, with the introduction of much less expensive, but far more efficient ICs, vastly expanded internal and external memories, and enhanced designs, even bistable ''storage'' has been tried and found wanting in comparison with today's constantly improving all digital designs. Does this mean that analog has expired forever? Probably not for now, but as technology further improves and certainly where only repetitive sampling is required, wouldn't you rather have the digital advantage for approximately the same price? Possibly not yet, but you might before completing the chapter. DSOs are extremely powerful and are very popular with folks who know.

Displayed waveforms also reconstruct from sample inputs at

prescribed intervals, but if the rate isn't 10 times higher than the frequency being investigated, you won't receive a full-bandwidth true sample. However, if this display is recurrent and not single-shot (aperiodic) then, over time, complete sampling will occur. Thus, already-sampling oscilloscopes are divided into two classes, and the one you want depends on which you need and are willing to pay for. If you don't deal in one-shots, then going the extra price isn't worth it. Between periodic and aperiodic lies equivalent time sampling, where scope sampling occurs sequentially at different points within the investigated interval. All of this work is done with trigger pulses, an internal ramp, and comparisons. When signal amplitudes are equal, an appropriate output delivers the reconstructed signal.

The actual digitizing effect occurs before waveform storage in solid-state memory. Initially, signals are digitally coded by an analog-to-digital converter (ADC) into digital bit patterns. They are then retrieved from memory, and the signal is converted and routed to CRT deflection plates for image display. A considerable advantage of the DSO is its ability to store and exhibit events prior to triggering. In analog scopes, the triggerpoint starts the horizontal beam sweep. In digital scopes, information before the trigger can be stored, reproduced, and analyzed. Further, any portion of the waveform can be read out motionless, so it can be studied in complete detail—something no analog scope can possibly do. This "freezing" can include both reproducing and expansion; waveshapes can be overlaid for comparison, if required or desired.

With this information now assimilated, we can say that the useful bandwidth (USB) of any digital oscilloscope amounts to:

$$\text{USB in MHz} = \frac{\text{digitizing rate in MHz}}{25 \text{ for dots and 10 for vector displays}}$$

There are further limitations, however, because true samples are only possible if they are at least 2 × that of the highest frequency, which is called the *Nyquist rate*. Usually used in working with digital data, it can also apply to analog information. Philips/Fluke warns, however, that instead of worrying about Nyquist, just be sure 10 samples per period are in any waveshape investigation.

The foregoing should be enough to get us started on the digitizing "road to glory" and make you feel comfortable with the explanations and applications to follow. Most should be fairly simple, but a few are positively difficult and deserve the best explanation possible—a task in which this author takes delight.

Technicalities

Vertical resolution is first. It can be judged by amplitude in volts, some percentage of full scale, or bit levels. The latter is a measure of the smallest volts value that can be digitized. For frequencies in excess of 15 MHz, Philips/Fluke suggests a 10:1 probe to relieve instrument loading. You are further warned that vertical resolution in percentage terms can be misleading because these are based on full-scale readouts, rather than smaller segments that are normally displayed.

Averaging can be a problem also when samples of repetitive signals are averaged and added to fill spaces between samples. This information is not all sampled; glitches, noise, fast signal changes, and single-shots are not included among the actual readouts. Ergo, loss of detail and even signal misrepresentation are possible. Averaging, however, in the form of quasipeak detection is often used by spectrum analyzers to smooth out selected patterns and reduce noise. So, all averaging isn't necessary bad, but its application must be specific and understood.

Display resolution is best realized by a "high-resolution analog-driven display," according to Philips/Fluke, thereby avoiding the usual problems that exist in economical unit displays. In large signals, however, a technique that is identified as "sine interpolation" is used for sine waves only to help reconstruct the original information. For other types of signals, "linear interpolation" is possible to identify actual and inserted signal additions.

Aliasing is a slower-than-normal sampling of waveshapes that results in probable distortion and lower frequency. This problem is hard to avoid, but it can only be rectified by observing Nyquist's rule of at least two samples per period. Bandwidth limiting is undesirable because signal frequency input is not only reduced, but it also skips fast waveform changes that are extremely important in both troubleshooting and initial design.

True, a "live" analog scope measurement is highly advantageous in capturing general information and many aberrations, but analog ramp horizontal deflections are usually not as accurate as those exhibited by the digitals, and nanosecond measurements plus very close tolerance information is considerably more accurate in DSOs. As signal-processing speeds increase, timing measurements become that much more critical and time differences, skew, cursors, digital readouts for amplitudes and frequency/time requirements become considerably more important. Some DSOs even have time domain reflectometry abilities, which are so very important in displaying cable and even PC-board faults, in certain applications. TDRs are not discussed in this chapter, but applied descriptions do appear in a later chapter that involves both copper and fiberoptics cable, as well as allied topics.

Below 10 nanoseconds, delays in analog channel mismatches are further problems and threshold signal levels set for basic counters and timers are subject to electronic drift. The DSO's time base is usually as accurate as those for counter/timers (possibly the same), so its utility is already established. One variety of DSOs will update measurements continually; this style is especially practical for signal monitoring over time.

Typical oscilloscope horizontal/vertical accuracies are ±3%, which is poor precision compared to representative DSOs. Many digital-storage oscilloscopes are quality-tested (by an IEEE Standard for Digitizing Waveform Recorders) with a ''pure'' sine wave of mathematically pure amplitude, frequency, and phase. Differences between this signal and that appearing on the DSO are computed as ''lost bits'' of resolution, as it supplies responses at particular frequencies. Most distortions are from signal generation, differential nonlinearities, and amplifier/digitizer noise. Therefore, only signal generators with maximum noise and distortion of – 102 dB are considered reasonable sources.

Glitch capture is becoming much more reliable (Fig. 4-2) as newer technology produces faster and better analyzers. At a rate of 100 megasamples (Ms/s), an ordinary DSO samples data every 10 ns. However, because stored data has to undergo processing

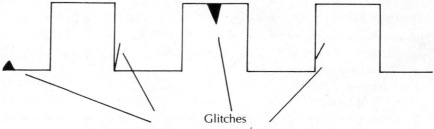

Glitches

Fig. 4-2 *Rectangular waves with accompanying glitches.*

before display, DSOs have "dead times" that are considerably longer than the analogs. As dead time increases, that much more information is lost.

To compensate for this problem, special acquisition modes have been developed, including a sequence mode that was instituted by LeCroy. In this mode, memory segments are "partitioned" into two-to-200 so that they can receive signal information during much or most of the dead time. Probably other such schemes are under consideration or already exist, of which we know nothing.

Horizontal loss can be captured via peak detection, which constantly tracks an incoming signal at slow-time base settings or via a process called *compaction*. However, peak detection, according to LeCroy, hasn't the necessary timing information, and problems arise with trace expansion. By maintaining a good sampling rate and extended memories, the waveform conceivably can be seen in its entirety. Compaction, however, is said to signify both the "position and size" of unusual glitches or transmissions and permit further signal expansion. Often, however, passive glitch-trackers operate only during the signal, so it might go unnoticed during dead times.

A special glitch trigger has now been introduced that operates independently of sampling rates. Once a threshold has been specified with timing, glitch identification and bit loss are probable. Occasionally, however, a glitch forces triggering, but it isn't shown on the CRT because it is outside the bit-stream detector. Another manufacturer has also developed a microwave frequency trigger for DSOs that's said to be effective to 20 GHz. Six band-pass filters are used and power levels are measured at the same time; the band that has the highest power level transmitted is accompanied by an AGC (automatic gain control) loop. Pin diode switches do the rest.

Resolution is the ability to resolve the information available—it is obviously extremely important. If sample rates and adequate memory is available, be positive that enough A/D conversion is available to see all of your subject. You must also consider instrument noise, nonlinearity, sample and holds and, in the end, effective overall resolution. Low resolution can be aided by averaging, but one-shot capture is something else again, so a choice will have to be made. However, extended bandwidths with repetitive characteristics can be acquired by repetitive sampling, which accumulates data following each trigger. This sampling requires time, of course, and delays are in both accumulating the waveform and displaying it. Just be sure of your DSO's capabilities and the tasks you'll require before attempting measurements on the bench or in the field. In other words, know your instrument! Furthermore, don't confuse resolution and accuracy. Accuracy involves worst-case input/output error, and resolution is the smallest voltage change that is fully represented by the action of the least significant bit (LSB). The 8-bit D/As appear fairly adequate for now.

A *quantizing error* (Fig. 4-3) is another error form as digital steps simulate analog levels. When any signal is digitized, offset and gain errors can be introduced anywhere in the DSO. Ground the input, for instance, and evaluate the output. If you don't read 0 volts, the output has an offset error. Integral and differential (remember your calculus) nonlinearities can also occur and these increase with higher frequencies. Harmonic and spurious distortion is next, followed by A/D additional nonlinearities. Try a few staircase signals, digital bit streams, and jittery signals for good measure, then check the competition. If you lean toward an instrument that uses Fast Fourier Transforms (FFT), low-frequency waveform accuracy is your guideline and the price can be more than considerable.

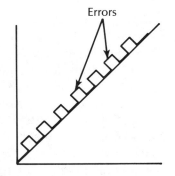

Fig. 4-3   *Digital conversion levels can be prime faults when they are nonlinear.*

In the physical/physics markets, further uses for DSOs (other than electronics) are also being discovered, which include strain, force, acceleration, pressure, displacement, torque, etc. Often this requires a simultaneous measurement of several signals and the ability to display transients, rather than the sampling of repetitive values. Frequencies are usually low, and 20 to 40 megasamples per second are often sufficient with transducers normally required.

Linear interpolation does permit the observation of risetimes and falltimes of visible waveforms, and it also indicates any resolution overload. Nonlinear interpolation is sometimes used to "extend" bandwidth beyond the usual divide-by-10 megasample-frequency rate. However, you had best have a long talk with the manufacturer's representative before attempting to acquire nonqualified instrument uses.

Finally, although 8-bit resolutions seem fairly standard for now, 10- and 12-bit A/D converters are even available for fine voltage resolution on low-level signals. In some cases, 8-bit flash converters operate better than 10-bitters. Once more, understand your instrument and know all requirements *before* going to the bank. Also, understand that four-channel DSOs don't have full range on the 3rd and 4th channels. These channels are set aside primarily for digital logic, and response is often quite limited.

Because of the remarkable storage abilities, accuracies, and minute-by-minute improvements of DSOs, what you usually see is about half of what you actually get. Therefore, we'll begin with Tektronix's newest, and work forward for the remainder of the chapter.

Tektronix' contribution

Although not specified to set the world on fire, Tektronix' 2232 (Fig. 4-4) is a nice, comfortable, easy-to-use box with an extensive set of instructions and a price tag that won't choke an undersized chipmunk. In addition, because it is both analog and storage at 100 MHz-MS/s rates, inexperienced storage engineers and technicians can easily make the transition between the two and also become thoroughly familiar with the significant advantages of each. Afterwards, if desire or need compel FFTs and/or super gigasampling (GHz), hock the local treasury and reach for the ultimate. Tek., LeCroy, and Nicolet will gladly accommodate with state-of-the-art performance.

But for a modestly-priced, thoroughly adaptable, reasonable passband instrument, the Model 2232 is a natural with sweep speeds to 50 ns/div. and vertical deflections from 2 mV to 5 V/div. A 10× magnifier, of course, increases the maximum sweep to 5 ns/div. (which should be fast enough), backed by a 14-kV CRT-accelerating potential. However, a good rule of thumb is to select an oscilloscope with a passband of 5× the frequency that you wish to measure. Consequently, 20 MHz is the correct fre-

Key Scope Controls

TRIGGER SECTION: Trigger controls allow you to select just the right time or event to trigger the scope—that is to tell the scope when to begin displaying data or in the Store mode, to acquire and display data.

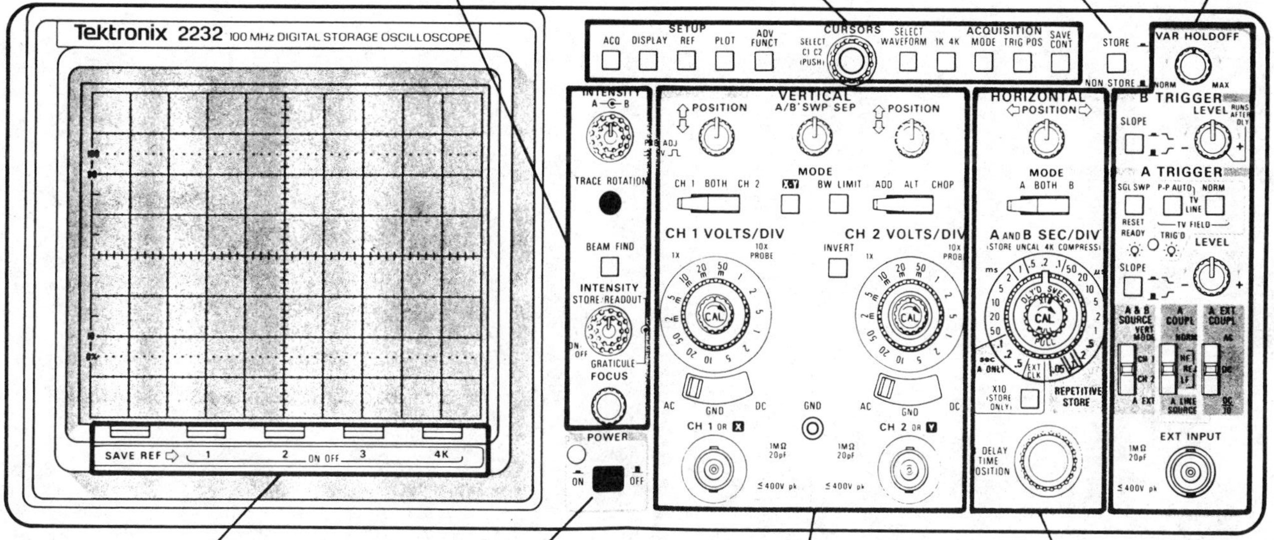

STORE/NON-STORE: Controls whether the scope functions in the analog mode or in the digital storage mode.

STORE SECTION: Controls menu displays, cursors, and methods of saving and displaying the stored waveforms.

CRT SECTION: Controls the brightness, focus, and alignment of the crt trace. Also, controls the brightness of the graticule lights. Provides a beam finder function to aid in locating displays deflected off screen.

HORIZONTAL SECTION: Controls the time scale (SEC/DIV) at which a signal is displayed. Controls the horizontal positioning of the display on the screen. In STORE mode, controls the rate at which the scope samples the input signal.

VERTICAL SECTION: Controls the vertical scale (VOLTS/DIV) of a displayed signal and its positioning on the screen. Provides input connectors and coupling for signals. Controls whether CH 1, CH 2, or both signals are displayed.

POWER: Turns scope on and off.

BEZEL BUTTONS: Control selection of menu items when a menu is displayed. In the non-menu mode, the buttons control the save, store, and recall functions for the reference memory.

Fig. 4-4 *Drawing with explanations of the Digital/Analog 2232 combination.* Tektronix, Inc.

quency bandpass for linearly accurate measurements. Above 20 MHz, some signals might be slightly degraded, but wholly usable to 100 MHz or better, which (some might recall), is the 3-dB downpoint that is recognized by the electronics industry.

For single-shot operation, the 100-MS/s rate looks pretty good because it can capture frequencies of 20 MHz or less. This rate translates to 100 display points per division, or a time base setting of 1 μs/div., which is a resolution of 10 ns. Vertical amplitude amounts to 2^8 (256) levels that are supplied by the A/D converter.

When setting up the instrument for single-shot capture, use peak detection and an arm for triggering, and set the trigger at some particular point, if you wish, on either the 1 024 or 4 096 selected records. Set up the menu, move the cursor to particular sample points for capture before the trigger (some experimenting is probably necessary). The result should be a very satisfactory, on-screen, frozen, one-shot that is suitable for photographing, X-Y analog plotting or dot-matrix printing. The ANSI/IEEE 488 GPIB and EIA Std. RS-232-C interfaces are available as options 10 and 12 for this particular DSO.

Approximately 10 points are required to characterize an event and you should have a 10 × sampling rate above the signal that you're measuring. Additional aid is also supplied by a trigger-level readout from the voltage.

Routine analog

On the analog side, you'll find the same two channels. Channel 2 is able to invert, alternate, chop (to 500 kHz), and convert the X-axis into a 2.5-MHz amplifier in the nonstore mode. Good for limited frequency phase matching, it can also be used for TV vector analysis when stimulated by a clean gated-rainbow color-bar generator. With adequate, short cabling and good-quality splitters, you can probably depend on the scope amplifier phase match to at least the high kilohertz. After that, the vertical amplifiers themselves will easily compare phased inputs into megahertz with considerable accuracy.

Time-base accuracy and trigger selections are especially noteworthy on this analog/DSO unit amouting to ± 2% in the nonstore mode and ± 0.1% in storage, with guaranteed delay jitter of 5 000:1. For triggering, sensitivities are 0.35 V/div at 10 MHz and 1.5 V/div. at 100 MHz. The trigger modes are p-p, TV line, auto, normal, single-sweep, TV field, 10:1 variable holdoff, and the usual dual A-B time base, which can quickly complement some of the storage activities when examining special time-related functions.

More information is available on this dual-function oscilloscope, but actions always transcend mere words, so let's demonstrate both analog and digital operations for proof.

The first waveform (Fig. 4-5) is a simple multiburst directly from a very accurate NTSC color bar and pattern generator. Used to evaluate video bandpass frequencies from 0.5 to 4.2 MHz, this range is illustrated by a color sync burst on the left, top display, followed by a 100 IRE (Institute of Radio Engineers, now the Institute of Electrical and Electronic Engineers) units, and the six periods of multiburst. By simply pulling out the time base B delay knob and turning a potentiometer control for an appropriate portion of the channel 1 output, 30.4 μs are delayed to plainly illustrate not only a small portion of burst, the horizontal sync pulse, the color burst (in detail), and part of the following "window," which precedes the 0.5 MHz start of the multiburst. Both displays are adjusted for 0.5 V/Div., the top trace is timed at 5 μs/div., and the lower expanded (delayed) trace is 2 μs/div. You can see very quickly how easy any waveform distortion could be spotted with such a dual analog, double-exposure technique.

Using only a single channel this time (Fig. 4-6) a second double exposure illustrates the difference between 10 and 1 μs/div. The NTSC pattern illustrates the six NTSC colors of red, blue, green, yellow, cyan, and magenta, which represent the so-called

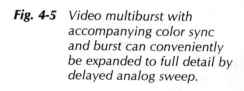

Fig. 4-5 *Video multiburst with accompanying color sync and burst can conveniently be expanded to full detail by delayed analog sweep.*

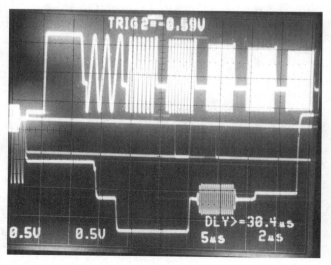

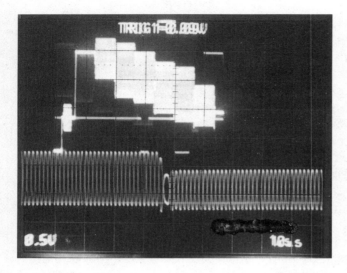

Fig. 4-6 *Double exposures: one at 10 μs/div. and the other at 1 μs/div. Both are NTSC color bar displays.*

''saturated'' colors in NTSC. As you can see, more than a considerable difference is between 10 × sweep times of the two photographs. Timing overlaps are deliberate to highlight the standard and delayed sweeps. This line of horizontal information is approximately 59.94 μs, of which 525 are in two fields or one frame.

Much of the remainder of the analog functions appear in the storage operation, where no problems occur with stable sync, frequencies, timing, amplitudes, and other phenomena that always make a storage oscilloscope so very useful. For once you're used to all the advantages of storage and its remarkable X-axis accuracy, it's a little difficult to return to pure analog for almost any close-tolerance measurement. Instead of painfully adding, dividing, or subtracting amplitudes and time or roughly calculating frequencies, all this work is done by marker and timing/amplitude readouts that you might find indispensable with continued familiarity.

Exciting storage

Digital scopes operate first by sampling discrete portions of any incoming signal and then transforming those samples into a binary number via a series of ascending levels developed by an analog-to-digital converter. A digital clock determines the rate of sampling that is delivered to the A/D, which quantizes analog information into individual binary levels. Such converters in intermediate storage scopes operate at 8 bits or 256 levels. Others are often capable of 12 or 16 bits, depending on the instrument. As you can understand, the accuracy of any A/D is especially important for digital resolution. Once such levels have been processed, they are stored in memory and read into a digital-to-analog converter for analog readout on the scope's cathode ray tube. In the 2232, the digitizing rate is expressed as megasamples/second (MS/s), or points/second, of which are 100 MS/s. An illustration of the simple storage controls is shown in Fig. 4-4.

Controls will be generally explained as we progress. Keep in mind that the release of the store button maintains the last stored waveform in memory; only one cursor moves at a time in any single-channel measurement, and both are active during the dual-channel save mode. In either the 1K or 4K storage modes are 100 data points/division or 50 data points when viewing two channels. In Triggered Scan, pretrigger data is updated between 0.1 s/div to 5 s/div or in the Ext. Clk Slow mode. Untriggered Scan updates the display continuously as each data point is received in P-P Auto. Setup buttons select the various menus, such as Acquisition, Display, Reference, Plot, or Advanced Functions (ADV Funct.) In Setup, as menus are displayed, Save Reference buttons at the bottom of the graticule select individual parameters, which are then viewable on the cathode ray tube. To commit this information to memory, the Sav Ref button needs to be engaged first, followed immediately by any one of the four reference buttons.

More of these significant instructions are contained in the 2232 portable oscilloscope operator's manual. Therefore, we won't belabor this discussion further. Simply let the on-screen Polaroid photos and captions speak for themselves, especially during unusual circumstances.

Figure 4-7 shows a simple sine wave with only the Storage commands in use, but it illustrates some of the complex readouts that can accompany. Delta V1 equals the amplitude of any of the sine waves from top to bottom. The negative trigger placement shows when the display is in the horizontal mode following the delay mode. The delta time measures total time between markers. Here, the 4K record length has been selected, and the bar graph points to the display position within the record; volts/div. are 0.5 V, amplitude is Average, and the time/div. is to 0.5 μs. If you would divide 0.5 μs into 1, for $F = 1/T$, you would discover an approximate frequency of 2 MHz. With 100 data points for each horizontal division, you can see that in the 4K (four kilobit) stor-

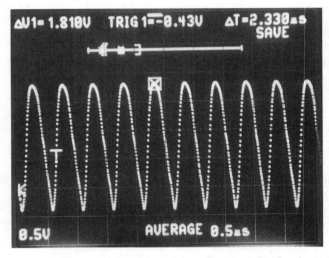

Fig. 4-7 *A simple sine wave in save and storage modes with amplitude, time, and trigger readouts.*

age mode, we're chugging along a little fast and many of the points are already showing. This is one of the singular reasons that more megasamples mean better resolution. These nonlinear sine waves (notice tops and bottoms as well as the right sides) could be much more closely traced if that was the case. The stored representation is very acceptable nonetheless.

Figure 4-8 is a different story altogether. For now, we're measuring the inverse of time, or frequency, with both markers at approximately the same no-difference zero levels. The *frequency difference* is now 0.588 2 MHz, the upper trace's amplitide is 1 V/div., and the lower trace is set for 2 V/div. The Mode is peak detect, the Delay is greater than 1 μs, and time bases are set at 0.5 μs. No trigger level is numerically apparent on the display, but it exists, as indicated by the "T."

Now, we're beginning to make real use of our storage settings, by gathering considerable information that can either be photographed (as shown) or outputted to a printer that has the proper format and cable connections. The Select Waveform button, but the way, selects either one or two waveforms, in addition to reference displays for cursor measurements.

Thus far, we've stuck to fundamentals that are extremely easy to use and set up; only scant attention has been paid to the Menu options at the bottom of the graticule. It's about time to illustrate some of these menus with appropriate explanations.

Figure 4-9 identifies the Acquisition mode calling for Trigger Position at 128 of 1K memory, in Peak Detect and Roll Slow. The "T" always denotes the trigger position. If you're happy or

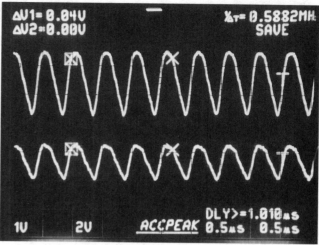

Fig. 4-8 *A single waveform, as seen through a 3.5-dB-loss splitter, at half the amplitude of the other. Notice the frequency markers at identical levels.*

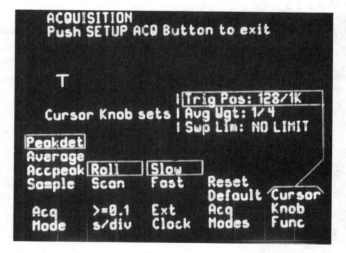

Fig. 4-9 *The process of establishing a menu for selective storage acquisition. Rectangles about each selection identify settings.*

otherwise with the selection, push the Setup Button to exit. Other nonrectangle-encircled positions could have been selected also, but everything can't be chosen at once. Now let's segue to Fig. 4-10 and see what happens.

Figure 4-10 is a composite dual storage on the scope's channel 1, which is easily accomplished by engaging the Save button twice. The first and larger display resulted from a false setting of the Trigger Level Control, which caused it to seek various pulse levels as erratic little steps, rather than a true display (illustrated in the lower waveform). Observe that the markers now record the

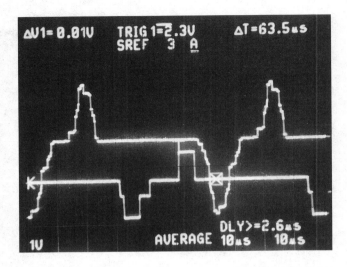

Fig. 4-10 *False trigger-level settings can result in many odd waveshapes as evidenced in the dual storage display.*

63.5 μs of a single cycle and also a Save Reference (SREF) of 3 on A. The readout is averaged and the time/division is 10 μs.

The moral of this oddball measurement is twofold: be sure of all trigger settings and remember that the Save Ref. programmer can select 1, 2, 3, or 4K to write whatever you wish into memory. You do not have to double-expose different selections. Simply push the Select Waveform button and the displays will appear and be counted. Amplitudes can be changed before the second Save, but the Time Base set should remain constant to avoid confusion. This, however, has been undertaken using only a single channel, other measurements could possibly engage both channels, provided that they operate on the same time base.

Figure 4-11 offers a better-than-usual example of sampling in a 6-step modulated staircase, which should really have been peak-detect. The trigger level rests at 1.14 V, the voltage difference between markers at the top and the lower sync pulse is 2.32 V, and the time separation between the top and bottom markers is 8.2 μs.

Although not a remarkable group of measurements, this photo illustrates many of the possibilities that are available in DSOs. In this instance, the sampling mode is distinctly not one of them. Peak detection would have been a smooth display that correctly illustrated individual staircase modulation.

With this particular digital-storage oscilloscope, you will easily discover many possibilities with use. You only need to become familiar with its readily accessible programming and try a few examples for both practice and applications. Most ques-

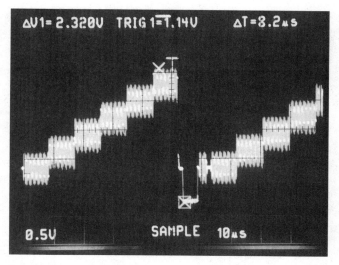

Fig. 4-11 *Sampling is plainly evident in this raggedy, modulated staircase. Peak detection would have been an improvement.*

tions can be readily answered by the 2232's operators' manual, which supplies a wealth of details and illustrations.

Hewlett-Packard

Hewlett-Packard is now making its mark in low-cost DSOs (models 54600A and 54601A) with two-channel and four-channel scopes, that have bandwidths of dc to 100 MHz and sampling rates of 20 MS/s. Channel 1 and 2 ranges extend from 2 mV to 5 V/div. Channels 3 and 4 are specified at 0.1 and 0.5 V. All four channels have an accuracy of ± 1.5%.

Both scopes are housed in a two-piece molded cabinet of tough, insulating plastic. The 7-inch raster CRT has a resolution of 255 vertical and 500 horizontal points (Fig. 4-12).

Eight bits of resolution produce record lengths of 2 000 points for single shots and 4 000 for standard sampling. In peak detect, the glitch capture is 50 ns per single channel and 100 ns for dual channel. Don't, by the way, be misled by 3rd- and 4th-channel specifications of 0.1 and 0.5 V/div. These specifications are actually recognized as trigger values and are fixed amplitudes of dc to ground. To expand such values by a factor of 10, use a 10 × probe, and increase the measurements to 1 and 5 volts each (suitable for standard logic inputs). Then again you've lost proportional sensitivity. Happily, the two main channels invert (triggering is also inverted at the same time) and we'll try and put them to interesting use beginning with applications.

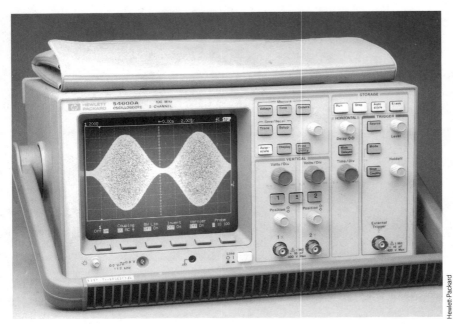

Fig. 4-12   *The HP 54600A digital storage oscilloscope.*

Displays include Normal, Peak Detect, or Averaging and trigger on Auto level, Auto, Normal, Single, and TV. A noise rejection feature is also included so that triggering will not occur (or will be less likely to occur) on noisy inputs. Bandwidth limiting protects against noisy signals, and it will not pass information wider than 20 MHz.

Although 1 × and 10 × probes are supplied with the instrument, a front-panel button selects both scale factors plus 100 × to match those of any similar probes that you might be using. As customary, capacitative probe compensation is needed to prevent overshoot/preshoot and low-frequency rectangular waveform rounding.

Weighing only 14 lbs., the 54600A is a true portable (small enough to be carried virtually anywhere, and it should fit comfortably under any airline passenger seat with at least 7 inches of clearance) and it is only 14 inches wide. The horizontal system supports sweeps of 5 seconds to 2 ns/div. with an accuracy of ± 0.01% and cursor accuracies between 0.01 and 0.2% of full scale, for both main and delayed sweeps. Internal trigger sensitivities are specified from 3.5 – 10 mV, depending on frequencies.

Measurements

Happily, the 54600A is fully automated with continuously updated measurements, including voltages that are averaged, rms, p-p, min., max., and timing for frequency, period, ± width, duty cycle, and rise and fall. Cursors are either automatically or manually positioned, and Autoscale sets vertical and horizontal deflections and trigger levels from 50 Hz up. An external trigger which ranges from 50 to 100 mV (± 18), is available on this DSO only, depending on the frequency inputs. The X-Y bandwidths are the same, but with a phase difference of ± 3° at 100 kHz.

To call special measurements or to institute specific commands are gray keys, soft keys, and action keys, which are white. Action keys, such as Run, Stop, and Erase, are the major commands. Gray keys inaugurate measurements, triggering, and channel, and they also set up the Soft keys under the CRT. These soft keys then do the final, definitive subprogramming characterizations like Coupling, Bandwidth Limiting, Vernier, Probe, Peak Detect, Average, Delta X, and so forth. And although this may sound somewhat complex, familiarity isn't at all difficult and soon becomes routine.

While we don't know the CRT accelerating voltage or its cathode drive, this is not an electromagnetically deflected cathode ray tube. It is an electrostatic one, much like those in early TV receivers, and its digitally-guided sweep rapidly transfers information to the tube so that plenty of brightness and detail is available for almost full daylight operation. Any reflections are only dimly apparent and indistinct.

The nine rotating front-panel controls are called *rotary pulse generators*, which actually have soft wipers that mate with PC board contacts. This arrangement is much different from those that use potentiometers, which are noisy, habitually collect grit, develop current burns, and soon wear out.

Glitches and single shots

These abberations are major reasons for today's advanced development of digital-storage oscilloscopes. A glitch, according to HP, appears as a very narrow waveshape that is considerably less than the main voltage. By combining good amplifier bandwidth with adequate sampling, Peak Detect and Display should help find the glitch once a stable display is on screen. OK for sweep speeds between 5 seconds and 50 μs/div., Peak Det. (Pk), but at sweeps greater than 50 μs/div., Pk does not operate and Autostore should be used.

Actually, HP says that Pk and Autostore can be operated jointly so that Pk registers the glitch and Autostore displays it in half-bright picture. Finally, you can also resolve the glitch with the Delayed soft key, expand the time base, and position this portion over the glitch with the Delay knob. Cursors or automatic measurements will do the rest.

Single-shot capture

For best results, try to gain some measurable impression of the subject waveform and develop a suitable trigger—especially for the waveform's leading edge. Engage a trigger source from among the soft keys, press Slope/Coupling, rotate the Level control for trigger placement, press Mode, then press the Single soft key. Erase prior measurements, and engage Run to arm the trigger. A second press of the Run key rearms it for another single event and erases the existing display. Comparison of several single-shots is possible using Autostore, which also arms the trigger and retains the image.

Single shots can be erased by any panel, soft key, or knob rotation. The Stop key, however, will reproduce the image and regain scope settings. Erase will clear the display, and either Run or Autostore will exit Autostore.

Remember that single shots have only a 2-MHz passband for a single channel and 1/2 that for dual channels, even though they capture information simultaneously. On the four-channel HP 54601, however, channels 3 and 4 capture jointly, immediately following channels 1 and 2 on the succeeding trigger.

In complex waveshapes, try to trigger on some sync pulse. When that fails, trigger on the middle of the waveform, and adjust the Holdoff for stable raster, even though the event is not synchronized with the trigger. Because this DSO procedure has nothing to do with time-base setting, you should be able to secure reasonable sync for both viewing and measurements.

Initial examples

Before working some of the more difficult measurements, let's undertake a few examples of how this equipment operates to understand its strengths and limitations. We'll also use the HP Think Jet printer to record.

First, you do not have to force the instrument into a storage mode to generate waveshapes that are suitable for a recurrent print readout. If you do, keep a fast finger handy to stop Autostore before the trace overexpands and distorts, because this function is actually accumulating storage, rather than one-shot digital.

Second, in the normal Run mode, all parameters can easily be set up so that useful measurements appear in the lower section of the graticule and also on the printout. As you will see, waveshapes are rapidly digitized by the equipment and are comfortably shown on both the CRT and on the print paper. Ordinary picture synchronization appears highly reliable and only some peculiar circumstances should or would cause problems. For the time being then, we will not attempt Autostorage. We will depend largely on regular digitized displays that are not actually stored. Our initial waveform is a good example.

In Fig. 4-13, you see a series of repetitive triangular waveshapes of approximately three cycles, with little top and bottom blips in between. It's obvious that we're accepting information into channel 1 between 7.812 and 107.2 mV, which the cursors measure as a delta channel 1 of 99.37 mV. In analog terms, you're reading amplitudes at 20 mV/div., with a 680 ms horizontal delay, a time measurement of 2 μs/div., and a 1:1 coupling, direct from the signal generator. What is not shown is a

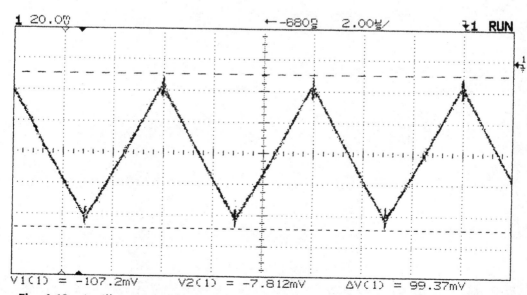

Fig. 4-13 *An illustration of how the HP 54600 measures amplitudes with cursors.*

– 58.75 mV trigger level, dc coupling, no bandwidth limiting, and no inversion. When satisfied with the display, all that's necessary, henceforth, consists only of pressing the Print command and the Software key. After a short pause, a duplicate display is dutifully printed. If excessive noise interferes, which it did not in this case, a suitable filter or bandwidth limitation will remove it.

The succeeding display in Fig. 4-14 appears as a rectangular voltage at a measured frequency between cursors as 122 kHz, and a duty cycle of 49.9%. Amplitude, the probe designation, remains at 1:1 because it's only a coupling cable, and receiver mode remains in Run. Observe that rising and falling edges are somewhat faint as frequencies increase. This is why faster sampling rates for specific measurements, such as one-shots, are desirable when operating under those conditions. This particular DSO will handle analog passbands to 2 MHz, so it is adequate for most waveshape investigations. Additional sampling, however, does not increase repetitive resolution because the scope is already undergoing maximum sampling activity and its subsystems will not accept more. Actually, when memory fills and the waveshape is recurrent, the scope is either operating over time at or near its analog bandwidth of 100 MHz anyway. This bandwidth is quite sufficient for most nonspecialized design and troubleshooting. What's more, your time base is now racing

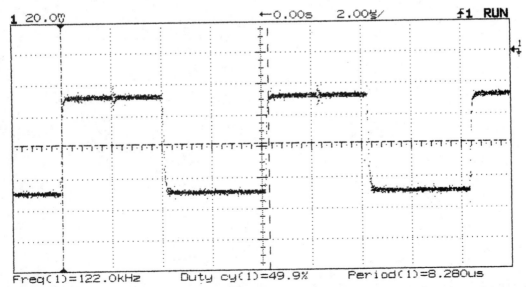

Fig. 4-14 *Periods, duty cycles, and frequencies are important when evaluating rectangular waves.*

along at 10 ns; therefore, considerable redesign would be necessary to accommodate both faster times and sweeps.

Glitches and one-shots, however, are another problem. In this case, the more gigasamples the better. When only momentary segments of a waveshape are available to collect, store, and reproduce, maximum aid is welcome under almost all circumstances. Therefore, let's see what can be done to simulate one or more of these annoyances and discover what the HP 54600A can do.

The One-Shot illustrated in Fig. 4-15 was set up with a very flexible pulse generator, that is adjustable over a considerable range at a fairly fast repetition rate. Amplitude scale on the DSO is set for 50 mV/div. and time base at 1 μs/div. Below, the actual waveform has a delta time of 280 ns and a frequency interval of 3.57 MHz. After jockeying the pulse so that the cursors could make rise and fall contacts, Autostore was engaged to enlarge the trace and present a better high-frequency outline, followed by a quick freeze-frame stop. Next, we went to the Menu and turned on Print/Utility, followed by soft key print screen. The results are very evident in a solidly-timed pulse that has an amplitude of approximately 60 mV and a nondisplayed trigger level of approximately 50%. This exercise should demonstrate rather conclusively that the HP 54600A can handle some pretty fast waveforms and deliver good, high-contrast readouts. Printer response, fol-

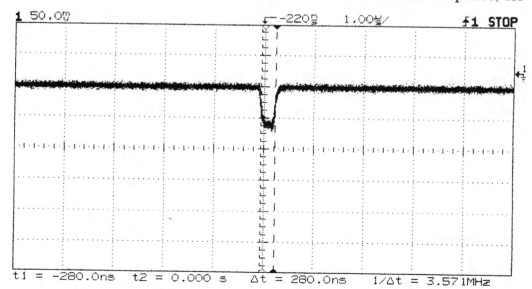

Fig. 4-15 *Measuring a negative-going one-shot is easy with a DSO.*

lowing hookup, is completely automatic and tear-off sheets are easily separated.

Finally, let's look for a glitch in some bit stream, which could cause trouble if left alone to plague some sensitive digitizer or decoder. The Glitch we're working with is just over 2 V in amplitude, and it appears in Fig. 4-16 just to the right of the ''O'' center vertical graticule line. Channel 2 (at the bottom) measures 8.480 μs between cursors at a rate of 117.9 kHz. The waveshape captured by channel 1 has a much slower repetition rate, but it certainly contains the elusive glitch. However, because the channel 1 and 2 traces are apparently related in frequency, the DSO can easily maintain sync so that both waveforms appear locked on the screen. Here, sync was taken from channel 2, which has sharp risetimes and falltimes—prized by any analog or storage scope. Observe that the glitch is approximately 0.7 V to the right of T2's ''risetime,'' which is actually a flip-flop cutting off and rising toward V_{cc}. So you'll have to discover whether risetimes or falltimes constitute the trigger (usually by the + or − slope) and adjust your incoming acquisition to determine the precise difficulty and troubleshoot accordingly.

Wrap up

The foregoing should offer some insight into the operation of digital oscilloscopes and why they are so very valuable. You'll find

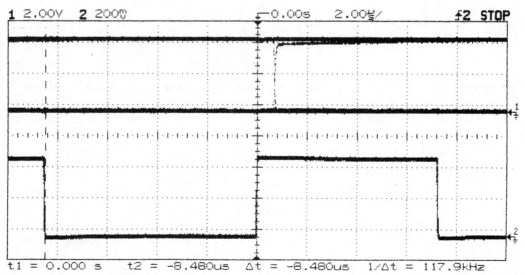

Fig. 4-16 *The extremely important glitch in the upper trace is not evident at lower repetition rates (below).*

that Autosetup is extremely handy, but when extraordinary measurements are required, the custom approach produces somewhat better results, even though it requires a little time to push and pull a few keys and knobs. It's also evident that digital storage has considerable advantage over analog, although often more expensive. Nonetheless, the digital approach not only presents a nonmoving display, but it also makes possible a very inexpensive paper printout that's permanent—even more so than photographs, which do fade in time. Moreover, an extremely accurate time base, programmability, and a good, bright on-screen display (for easy viewing) are all recommendations.

The HP 54600A and its four-trace companion, the 54601, do have a significant advantage over much of their competition because all acquisition and display functions are entirely separate from the central microprocessor, a Motorola 68000. A pair of custom ICs have relieved the microprocessor of these chores, so it has now developed update display rates that are equal to analog oscilloscopes, placing on-screen information before its operator very rapidly.

The block diagram that appears in Fig. 4-17 offers a useful perspective on these advanced operations. At the bottom, you see the Processor Bus, which accepts or delivers feeds to/from the Keyboard, the Option Interface, the Analog Controls, the CPU, and the pair of ROMs and RAMs.

Signals that enter channels 1 or 2 are usually attenuated, preamplified, and multiplexed for triggering. They are simultaneously supplied to Track-and-Hold circuits that feed Analog Converters and the Acquisition Processor for the Waveform Bus. This bus delivers and retrieves data from the Waveform Memory, Translator, and Video RAM, which energize the display.

Knowing that the old sampling must be erased before new points are registered, HP discovered that addition of the two acquisition and display ICs would not only accelerate data registry, but they would also relieve overtaxed microprocessors from delivering grainy images and tendencies toward Nyquist limits, which cause slow-frequency aliasing and even distortions. The HP 54600 series can now contribute 4 000-point record lengths with excellent sampling and better-than-average digital displays at both slow and high frequencies.

Briefly, the Acquisition Processor is a 1-μm CMOS with two datapaths under control of a pair of programmable-logic arrays, and a 2K $\times$ 8 random-access memory (RAM). Triggered at 100

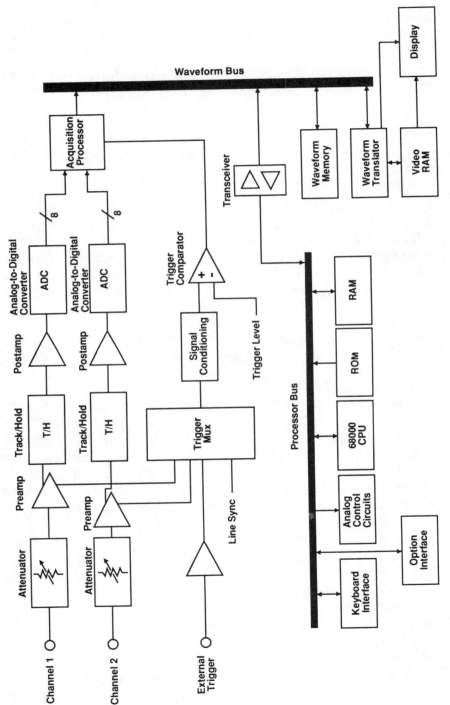

Fig. 4-17 *A functional block diagram of the HP 54600 illustrates overall operation.* Hewlett-Packard

MHz, most of the IC operates at 20 MHz as it drives track-and-hold circuits, two A-to-D converters, an interpolator, as well as additional counters and adders and dedicated logic, which is written asynchronously into the RAM.

The Waveform Translator contains a 1.5-μm gate array that operates at 40 MHz to excite a 256K × 4 video DRAM (dynamic random-access memory). This portion of the IC divides into four pixels, two for full bright display and two for half-bright display. The latter is used for text, stored waveforms, and other allied images. Overall, the RAM itself consists of eight 4K-byte waveform voltage data information. As live waveforms enter the dim-bright displays, an additional operation, named *Autostore*, occurs. Information plots into both full- and half-bright video images, accumulating in intensity until frozen by the Stop button. Thereafter, the Translator can erase pixels, generate cursors, and deliver special access modes to the video RAM via the microprocessor. An erase cycle occurs at the conclusion of each video frame.

Although we did not operate this DSO with a computer, nor did we record special pretrigger activity, both types of storage and display are always available and should be used whenever necessary. Noise-rejection features were engaged and proven effective in a number of instances. The X-Y plotting, together with a 10-gated rainbow color-bar generator produced a fair reproduction when connected to the red/blue cathodes of a color television picture tube, with both channels inverted to deliver a correctly phased pattern. We would have to admit, however, that a good analog reproduction is somewhat more satisfactory, because it does produce more detail and a smoother image. On the other hand, multiburst oscillations from 0.5 – 4.2 MHz were faithfully reproduced in usable detail.

As digitizing rates and expanded storage increase, however, some manufacturers will depart from the analog scope business altogether, depending on the utility and flexibility of a well-designed DSO. Others will remain half-and-half for the time being, while diehards cling to low-cost analogs for a dimishing portion of the market.

In the meantime, all three categories are expected to improve, but as the cost and performance curves contract and expand, the DSO is certainly headed for a significant piece of instrument sales and usage. At the moment, 100 MHz and 100 MS/s seem about the right mix for general applications, even though they're somewhat more expensive than the 20 MS/s group.

As for the dual analog and storage-scope combinations, many engineers and technicians still like the smoothness of analog processing, but with the storage and display capabilities of the DSOs. At the moment, we can't guess the market directions, but publicity, education, applications, and engineering developments certainly are encouraging for all worthwhile DSOs.

Troubleshooting

One further item is the ability of this instrument to troubleshoot itself. Many instruments will have to duplicate this development in one form or another as solid state becomes even more compact and extensive. Little probes and even the best digital oscilloscopes won't do the job on a 60-pin microprocessor, or even on a group of 24-pin logic gates. Some sort of internal stimulation and prompting will have to occur to reassure readouts, otherwise, you'll never know if on-screen displays are even partially accurate. HP, therefore, is to be generously commended for this sorely needed attention to detail, which we'll briefly present as an excellent example of operational checks and specific examinations. As such, these examples are basically subsystem reads, but they decidedly serve to warn of deeper malfunctions or outright assurances that "all's well."

Ever hear of a DSO with worthwhile troubleshooting instructions in the User and Service Guide Manual? Well, this DSO has one! The only external equipment required would be an oscilloscope, a digital multimeter, a probe, and a dummy load—if you are especially serious, that is.

Otherwise turn off power, disconnect signal cabling and any interface modules, then engage the power line (mains) again. All commands now appear on the CRT as rectangles and the knobs as circles. As each key is pressed, the rectangles fill with greenish light and the knobs show arrows in the directions that they're turned. If all keys and knobs operate, the keyboard and its assembly doesn't need to be replaced (Fig. 4-18).

The Digital-to-Analog Converter (DAC) is next examined by first connecting a multimeter to the dc Calibrator. Press the DAC soft key. Your multimeter should read $0 \pm 500 \ \mu V$. Any other key will cause the meter to display $5 \ V \pm 10 \ mV$. If all's well, you do not have to replace the system board.

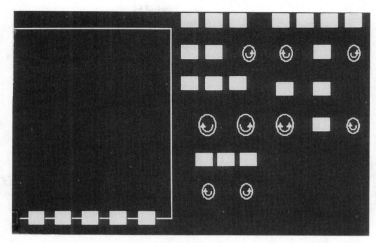

Fig. 4-18   *A CRT image of troubleshooting a DSO. The rectangles represent the keys and the encircles arrows represent knobs.*

The read-only memory (ROM) is next, and it simply operates by pressing the ROM soft key. If the readout displays *test passes,* your ROM works. If not, and *test failed* appears, you must replace the system board.

Then, is the random-access memory (RAM). Simply engage the RAM soft key and look for *test passed* or *test failed.* If the latter appears, replace the system board.

Last is the Display Test. It's another simple test that engages the Print/Utility button, the Self test soft key, and the Display soft key. If half-bright and bright squares materialize, the display is good. Otherwise, replace it. Any additional key pressed causes squares and borders to appear in all four CRT corners. If not, the system board needs to be replaced. Any key (not a soft key below the bezel) pressed then completes the test.

Is that simple enough? These are absolutely the simplest and most effective tests imaginable. Our compliments to all of the HP engineers and programmers involved.

❖ 5
Our wonderful world of spectrum analyzers

A HIGH/LOW-FREQUENCY INSTRUMENT THAT'S OFTEN CONFUSED with its nephew, the oscilloscope, the spectrum analyzer will tune and detect electronic signals from dc to medium gigahertz (GHz), analyze their content in terms of frequency, time, decibels (or linearly), and display the result with remarkable accuracy and detail. As you should already perceive, scopes operate in voltages along the Y vertical axis, and in appropriate divisions of time along the X axis. So, they seem to be distant cousins, rather than entities of the immediate family. However, one often supports the other as a result of voltage/dB and time/frequency convertabilities in the general sense. However, they're worlds apart in terms of signal sensitivities, frequency spreads, image portrayals, internal electronics, and especially costs.

A modest oscilloscope can be had, spanking new, for under $1,000, an honest-to-goodness spectrum analyzer with useful features is formidably priced over $5,000 and up to $35 and $100 thousand dollars, depending on the accuracy and frequency range. Although most oscilloscopes do the same tasks as their brethren, analyzers are often highly specialized, depending on range and readout, and they don't always supplant one another in specific tasks. Therefore, in choosing a spectrum analyzer, it's very wise to precisely know your requirements and purchase accordingly. Don't avoid the inevitable sales talk, but make it work for you in accumulating information and examples. As this chapter proceeds, we'll try and demonstrate exactly what's meant by both discussions and demonstrations from at least audio to geosynchronous satellite downlinks at 4 and 12 GHz—an almost dc-to-10^9-Hz frequency spread.

Analyzer topics and selections

Usually, types include phase-lock, microprocessor-controlled, synthesized, and a new group of low-cost analyzers with manually-dialed center frequency readouts that are primarily useful for general displays, rather than as accurate, wide-ranging instruments. Fast-Fourier Transform (FFT) equipment is also included, but most of these are very expensive and execute relatively low-frequency tasks with considerable accuracy. Later, FFT abilities might change, but for now, kilohertz and low megahertz ranges are maximum. A few analyzers are also multichannel and possess built-in zoom features for closeups, but, again, they are usually low-frequency equipment that are quite costly. Nonetheless, the spectrum analyzer price trend is downward and many new models will be available during the 1990s with more features and better values than we've ever known in the past. Meanwhile, many varieties are available at a multitude of values to do the job if you have carefully evaluated your requirements.

In chapter 1, we discussed various types of oscilloscopes and their outstanding features, therefore applications throughout this chapter will take precedence over further individual discussions so that you can derive maximum benefit.

In making your selection, however, beware of instrument sizes and weights. Portable units are both desirable and advantageous, but some of them weigh a ton. So, it's best to read a specification sheet before passing the ducats and straining your back. In addition, self-calibrating equipment is extremely important because of shop charges for this purpose are considerable and can be easily accomplished for free in the better units. Specialized software routines, such as operator-processed or card-inserted programs, are also extremely important if your particular task warrants it, now or later. This approach should obviate the need for expensive, dedicated plug-ins to which the industry has become accustomed. Naturally, the analyzer system has to be microprocessor or microcomputer-controlled to accept special changes and applications.

Well-lit graticules are also important if you take CRT-trace pictures and expect full grid and waveform displays. For one-shot captures, well-lit graticules are indispensable. Generally speaking, with no lit graticule, and no X/Y time divisions showing, the measurements are always inconclusive.

Bright CRT traces and tiny spot sizes are also determinants of worthwhile definition for both on-screen photos and outdoor operation, usually required by cable and satellite troubleshooters and installers. For 5-inch tubes, 14 to 16 kV is probably the minimum requirement. Otherwise, you'll need a viewing hood, which would limit general operations considerably.

Several storage registers should be available for multiple readouts when related signals need to be seen and compared. Menu-driven instruments are much preferred for fixed parameters, because they're programmable and often considerably more accurate.

LEDs help identify which amplifier or trigger remains in use, and whenever you wish to have dual-channel amplifier identification, as well as amplifier addition or special triggering. Where special menus are involved, lamp identification is always thoroughly helpful and is often mandatory. You'll need LEDs, too, for peak detection, markers, graticule lighting, sweep timing, signal positioning, and other functions that are not already read-out on the CRT screen.

Maximum sensitivities in excess of − 100 dBm are always welcome—as long as the total operating window is at least 70 dB and the analyzer's top range approaches or exceeds + 30 dBm. If not, you have a limited-range instrument that's often suitable for only restricted measurements.

Guarded inputs for both 50- and 75-Ω impedances are necessary if you aren't willing to add a dc-blocking capacitor in series with incoming voltages. Otherwise, the even-Stephen chance is that you'll overload the first local oscillator (Lo) or amplifier and pay a considerable bill for reconstruction/replacement and full equipment calibration. A protected input of about 100 Vdc is suggested for general usage.

Special setups for noise, bandwidth, and carrier-to-noise are desirable, as well as any available markers. These are needed for specific measurements, involving video, satellite and other higher frequency bandwidths that are not related to lower frequency measurements. Noise, for instance, could be 1 Hz, video 4.2 MHz, and BW at − 6 dB down.

Tracking generators are usually no more accurate than the spectrum analyzer's local oscillator (Lo). Although useful for filters, amplifier characteristics, and possibly standing wave ratios (SWRs) with the addition of a return loss bridge, they are expensive additions and not required for the ordinary measurements.

Quasi-peak detectors are great for those engaged in radiation testing and will smooth-out displays required in FCC Part 15 regulations. They're actually three filters, and the detectors are active in measuring electromagnetic interference (EMI).

Synthesized analyzers have master references that are digitally transmitted throughout the instrument. Although usually much more expensive than either phase-locked or microprocessor-controlled instruments, their accuracy is considerably greater than nonsynthesized equipments.

Fast-Fourier Transforms are illustrated mathematically as an integral from minus infinity to plus infinity and calculate power as a function of frequency. Any waveform, then, becomes a mathematical approximation. Periodic, as well as nonperiodic excursions, can be represented as finite dimensions over an exact interval. Difficult to design and probably to manufacture, they do offer fast, high-resolution images for dynamic signals, especially, and they measure all frequencies at once, rather than individually. Because they measure both the magnitude and phase of the various frequency components, they are useful in signal statistics, as well as the relationship between several signals. Normally lower frequency units, a few manufacturers make these analyzers and others might enter the market if specialty sales increase.

A few other specialties and suggestions do exist, of course, but these are the prime considerations and are the ones to which we'll be paying special attention throughout the chapter. We'll specifically cover the manually-tuned group with fairly restricted ranges, the intermediate equipment that is phase-locked and microprocessor-controlled, and at least one synthesized instrument that has both high accuracy and high frequency. In this way, not only are instruments and their applications described, but also most equipment is categorized in terms of ability and utility. We'll start near the bottom and work up.

Modestly-priced equipment

Let's begin with simple, center-tuned, rather restricted units at modest cost and work up, expanding the effectiveness of each from one application to another until we're well into the GHz spectrum (home of microwaves, radar, and satellites).

Two notable low-cost units are now available at prices that were unheard of in the 1980s. One, a 500-MHz Hameg analyzer (Fig. 5-1) stand-alone unit that operates between 500 kHz and 500 MHz was initially priced at $1690. Its center-frequency accuracy lists at ± 100 kHz and marker accuracy at ±0.1% of span

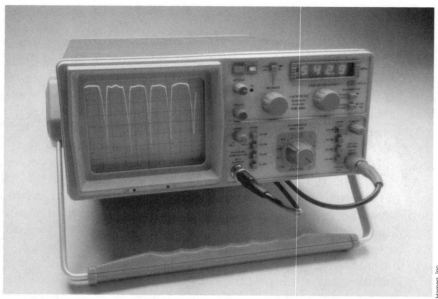

Hameg, Inc.

Fig. 5-1 *Hameg's recently introduced stand-alone 500-kHz to 500-MHz analyzer, complete with tracking generator.*

plus 100 kHz, with frequency stability of less than 150 kHz/hour and bandwidth resolution from 250 to 12.5 kHz, and a scan rate of 43 Hz.

Amplitude range extends from − 100 to + 10 dBm at 50-Ω input impedance, and the log-scale accuracy is ± 2 dB. The scope weighs 17 pounds and has a 6-inch cathode ray tube. Also featuring a tracking generator, the HM 5005 specifies this added equipment at an output level of − 50 to + 1 dBm in 10-dB steps and an adjustable output attenuator of 0 to 40 dB in four steps, with a frequency range of 100 kHz to 500 MHz.

A second, recent introduction is BK Precision's PL 5610 ac/dc portable analyzer (Fig. 5-2), which offers a bandspread from 1 to 1 000 MHz, a 3.5-digit LCD display with 3-MHz calibrated accuracy and 6% center-frequency accuracy over 100 MHz. Its horizontal scan time measures 3 ms (adjustable 5%), and amplitude measurements range from 15 to 123 dB μV and it is dc-guarded to 50 V. Above 10 MHz, the dynamic range lists at 70 dB (10 dB per division) at an accuracy of 2 dB, with a switchable input impedance of either 50 or 75 Ω. Totally portable with a convenient carrying handle, battery power lasts for two hours after recharging, it operates from 100 to 230 Vac, and current require-

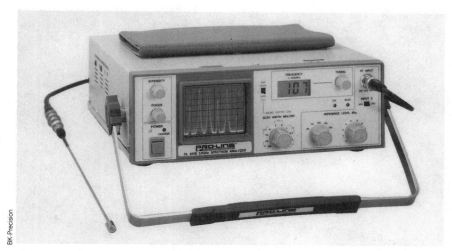

BK-Precision

Fig. 5-2 *The equally new BK-Precision low-cost analyzer with center-frequency LCD readout and bandspread from 1 MHz to 1 GHz.*

ments are approximately 1.2 amperes. The quoted user price is a modest $2995. Calibrated in microvolts, a table is included for cross-referencing to dBm. Resolution bandwidths are adjustable from 10 kHz to 1 MHz by front-panel switching. The CRT measures 3.5 inches, with 8 vertical and 10 horizontal divisions. The PL 5610 weighs 16.5 lbs.

Depending on requirements, these two are worthwhile inexpensive units with fair accuracy, limited ranges, and limited resolution bandwidths. If frequently hand-calibrated (the procedure is simple), either will serve many purposes within their several abilities. Recall, however, that audio frequencies begin at about 30 Hz and expand to at least 20 kHz; AM radio broadcasts are roughly 500 kHz to 1.7 MHz; TV broadcasts extend from 54 MHz to 806 MHz (except for FM at 88 to 108 MHz). Any of these carriers, though, are power-related and measured only in dBW or dBm, the difference is 30 dB. Neither of these analyzers can perform in satellite frequencies following block downconverters.

The next step up is the R4131D by Advantest (Fig. 5-3), a Japanese manufacturer with U.S. headquarters in Illinois, which sold originally for $5,995. Weighing 22 lbs. and with a wide-frequency range from 10 kHz to 3.5 GHz, this analyzer has a maximum input sensitivity of – 116 dBm, a dynamic range of 70 dB, AFC control for oscillator stability, a save/recall function, video filters from 10 Hz to 300 kHz, a 5.5-inch CRT with amber display, center-frequency accuracy of 10 MHz, resolution bandwidths

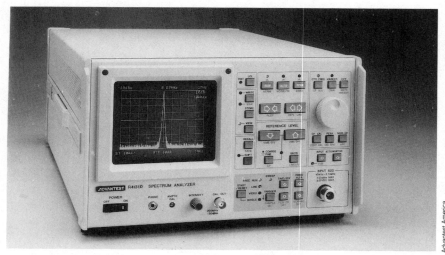

Advantest America

Fig. 5-3   *Another lower-priced analyzer mod. 4131D with exceptional pass-band of 10 kHz to 3.5 GHz.*

from 1 kHz to 1 MHz, and an RF attenuator setting from 0 to 50 dB in 10 steps plus maximum hold and peak search. This price is reasonable for an analyzer with such a range and recorder/video outputs, and a GPIB bus.

Third in line, price, and features, the Anritsu's MS610B/J (Fig. 5-4) has an announced price of $6,995. This 27-lb. unit has a frequency range between 10 kHz and 2 GHz, hardcopy readout with XYZ inputs, resolution bandwidth from 1 kHz to 1 MHz, 10 dB and 2 dB reference levels, 100 Hz and 10 kHz video filters, input attenuator 0 to 50 dBm in 10 dB steps, sweep times from 10 ms to 99 s, a calibrator output 50 MHz 6-inch P39 long-persistence phosphor CRT, a dynamic range of – 50 to + 20 dBm, and dynamic-range residual response beginning at less than – 100 dBm.

Next is certainly the IFR model A-7550 at $7,195 (Fig. 5-5), which has a frequency range from 10 kHz to 1 GHz, a 70-dB dynamic range, automatic calibration, is fully RF-synthesized, has 300-Hz resolution bandwidth, menu-driven displays with keyboard entry, log/linear readouts, digital storage, average and peak-hold modes, 300 Hz to 30 kHz video filters, a 7-inch cathode ray tube, all common-amplitude units, inputs from – 120 to + 30 dBm (dc-guarded to + 4 volts), and input impedances of either 50 or 75 Ω with adapter. A tracking generator is optional.

Fifth in the more modest price grouping is *Tektronix' model*

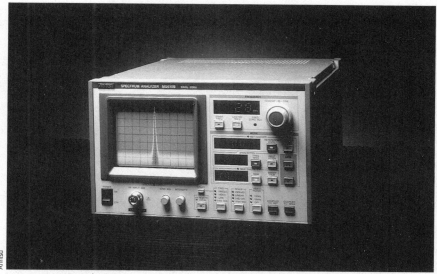

Fig. 5-4 *A second Japanese analyzer with high dynamic and frequency ranges and GPIB equipped.*

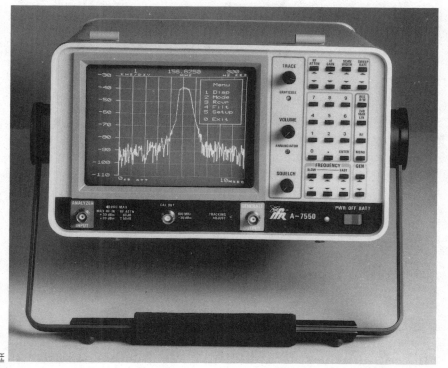

Fig. 5-5 *IFR's menu-driven, 7" CRT with frequency ranges from 10 kHz to 1 GHz.*

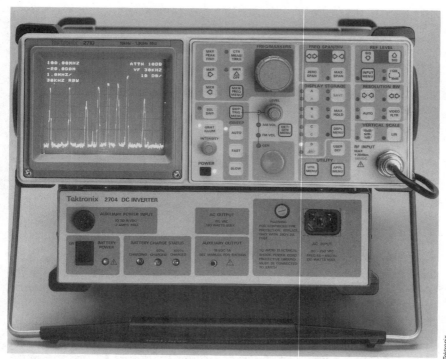

Fig. 5-6   *Tek's versatile and famous 2710, which is phase-locked and menu-driven with a frequency range from 10 kHz to 1.8 GHz.*

2710 (Fig. 5-6) with which I have personally worked for two years. This highly developed, self-calibrating, microprocessor-controlled, phase-locked equipment is fully menu-driven, with a 1×10^{-5} center-frequency accuracy of ± 5 kHz, 50- and 75-Ω inputs guarded to 100 Vdc, dual reference and frequency markers, displays at 10, 5, and 1 dB/division (plus linear), with accuracy at 10 dB/div over the 80-dB range (cumulative) of ± 2 dB (better in other ranges), and at 3 kHz with preamplifier sensitivity amounts to -129 dBm. Sweep rates are 1 μs to 2 sec/div., with instrument settings, waveforms, and other routines stored in nonvolatile RAM with battery backup. Internal calibrator drift is ± 10 ppm over one year. Resolution bandwidths extend from 3 kHz to 5 MHz, it weighs less than 25 lbs., and it includes a 5-inch cathode ray tube with an accelerating voltage of 14.5 kV and 7 different multiple-selection menus, including carrier-to-noise, with AM and FM detectors, video display, and printer/plotter facilities as options. The zero span operates from 10 kHz to 100 MHz.

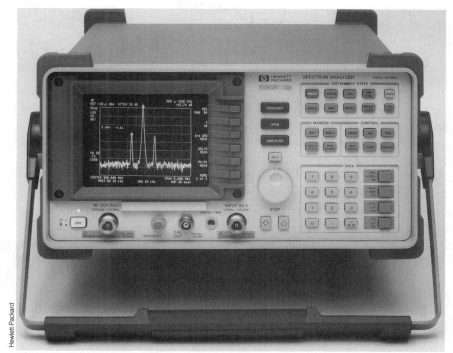

Fig. 5-7 *The HP 8590B, also phase-locked, ranges from 9 kHz to 1.8 GHz with good sensitivity.*

The Hewlett-Packard HP 8590B (Fig. 5-7) is the final unit in this intermediate series. The HP 8590B costs just under $10,000 at $9,590, and the two new HP HP8591 and HP 8594A analyzers at approximately $13,000 and $15,000 (respectively), each of which has additional features, with the HP8594A extending its top range to 2.9 GHz. The basic unit, however, is the HP 8590B and on this one we'll concentrate.

The HP 8590B is a new addition to HP's analyzer line. It's phase-locked, has a frequency range from 9 kHz to 1.8 GHz and sensitivity from – 115 to + 30 dBm, and in 0 span the range is from 10 kHz to 1.8 GHz. Inputs can be either 50 or 75 Ω (as customer specified), it has 10- and 1-dB log scales (plus linear), and a frequency response of ±1.5 dB at 300-MHz cal. output, with frequency accuracy at ±5 MHz.

Sweep times are from 20 ms to 100 s and with 3% accuracy, the weight is 30 lbs. An optional card reader allows the loading of dedicated programs, and an optional built-in tracking generator allows synchronously swept signals. Calibration here is fully

automatic, and internal memory stores 50 signal traces. Printer/
plotter option outputs are also available.

The analyzer priced next to HP's HP8594A would probably
be the Tektronix model 492PGM (Fig. 5-8), which has been selling for $20,900. Reputably a portable, but weighing 46 lbs., this
instrument is more for the laboratory, unless you really need its
remarkable specifications in the field. Completely synthesized
with a standard frequency range from 10 kHz to 21 GHz, its frequency accuracy is ±30 kHz at 1 GHz with frequency accuracy of
$1 \times 10^{-5/year}$. The dynamic viewable range is 90 dB and sensitivity is −134 dBm, full IEEE-488 (GPIB) programmability with Tek
codes and formats for standard bus operation, markers, extensive
on-screen readouts, and occupied bandwidth and noise normal-
izations via pushbuttons. The nonvolatile memory accepts up to
9 waveforms and 10 front-panel settings, resolution bandwidth is
from 1 kHz to 3 MHz, sweep range is 10 ms to 20 sec/div. at a
±5% time accuracy, and the maximum input power is 1 W (+ 30
dBm). Dynamic markers automatically update frequency and
amplitudes with each sweep and external mixers can increase the
frequency range to 325 GHz.

This latter inclusion was inserted deliberately to illustrate
the difference that $10,000 makes in first-class instrumentation.

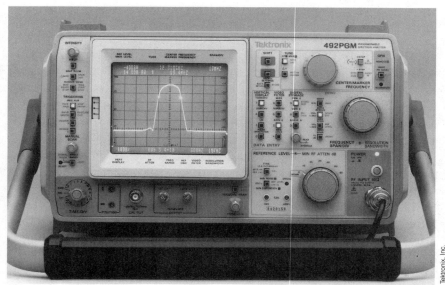

Fig. 5-8 *Tek's 492 PGM, is somewhat more expensive, but it has a range
from 10 kHz to 21 GHz, is versatile, and is completely synthesized
for maximum accuracy.*

If you need such specifications, then there's no substitute for quality, even though the cash outlay might often be difficult.

Although we haven't covered some of the more remarkable and very expensive analyzer equipment available, the foregoing will offer an idea of how your needs can be fulfilled to one degree or another with analyzers now on the market, and it can serve as something of a forecast of what's to come. In that regard, late 1991 will see a number of new analyzer introductions at even more competitive prices and additional features to make them more popular than ever. Now, let's see what these instruments can do . . . .

Analyzer applications

Recall that spectrum analyzers are not necessarily very low frequency instruments, unless specifically designed. Many of these are fast-Fourier transform types for high resolution, log and special sine-wave analysis, networks and phase, as well as linear or log sweeps with increasing or decreasing frequency, and selectable sweep rates and resolution. Accuracies, such as 0.15 dB, are extremely high, inputs are often dual channels and Hewlett-Packard's HP3563A can accommodate digital inputs up to 16 bits wide in parallel and two's complement. So, for these unique measurements, FFT analyzers are extremely important.

For our purposes, however, we'll stick to the broad-frequency coverage swept-tuned superheterodyne receiver, often described as a frequency-selective, peak-reading voltmeter. Operating in the frequency domain primarily, with frequency, maximum, and zero span. In frequency span, the analyzer is operating normally at whatever sweep is required within limitations of the instrument. Maximum span denotes the single maximum sweep span available and is fixed. However, zero span remains only at a designated center frequency with no overall sweep.

In *zero span*, the resolution bandwidth actually brackets the selected center frequency. With that portion still swept, the readout shows time/division (not frequency), so your analyzer is now a limited range oscilloscope in either dB or linear modes. Because the swept-tuned superhets are the primary instruments in the analyzer category, we'll work with them.

Signal-to-noise (terrestrial and satellite)

Unfortunately, many technicians and even fairly experienced engineers remain either confused or ambivalent over *signal-to-*

noise (S/N). Simply stated, virtually any reasonable display has more "live" or working content than noise. Therefore, the signal portion is of major interest. Noise, however little, can sometimes become a problem. This is especially true when noise rises toward the signal or when the signal sinks into noise—either way, when relatively few dB separate the two, the ratio of signal-to-noise is a real difficulty.

Noise, for instance, can be generated thermally (heat) as a result of certain frequencies and rapid current flow in semiconductors, intermodulation products, signal cross coupling, noisy oscillators, external circuits or system noise, switching power supplies, ground loop or circulating currents, sky/ground noise, and more. In simple baseband video applications, any measurement below 35 dB shows visible snow (noise).

Outdistancing S/N in many instances, however, is carrier-to-noise C/N because of a direct relationship between the two. Whenever an electromagnetic signal rushes through the sky or air, a certain amount of noise caused by distance (signal fading), external sources, electrical phenomena, etc., is received, joining the transmission. When this signal reaches its destination, antenna, cable, amplifier, and even external interference add to the overall problem and they become a part of the undesirable disorder.

As the frequency rises, especially to the microwave region (where satellites, radar and allied receive/transmitters are operating), noise becomes critical because of the tiny signals that are available on carriers. In the 4- and 12-GHz C and Ku downlink bands, for instance, a net transponder C/N of 12 dB will result in pretty fair video, because a C/N of 12 dB would mean:

$$S/N(FM) = C/N + 37.5 \text{ dB}$$
$$= 49.5 \text{ dB}$$

and studio quality for satellite S/N is 56 dB. Satellite transmissions, of course, are either frequency modulated or will eventually become digitized, either partially or altogether. Television broadcasts, on the other hand, have amplitude modulated video and their C/N should certainly be above 40 dB, because:

$$S/N(AM) = C/N + 5 \text{ dB}$$
$$= 40 + 5 \text{ dB}$$
$$= 45 \text{ dB}$$

and studio quality is 54 dB. Just so you'll have a "feel" for transmissions terrestrial and in space, a few dB is all we have to worry about in normal TV signal areas, but from space, you lose 196 dB for C band and 206 dB at Ku. Also, signal levels for either satellite propagation following block downconverters or TV broadcasts must reach levels approaching 40 dBm (power measurement) for adequate pictures.

Let's look at some terrestrial and satellite examples to prove the points. In addition, you'll learn how to cope with analyzers that do not have automatic C/N measurement compensation. This point is extremely important as a result of the additional built-in losses. First comes TV broadcast signals, followed by those from geosynchronous satellites 22.3 kilomiles in orbit above earth's equator, where they are positioned and station-kept within 0.5 to 1°.

Let's work with a nonautomatic 7L12 Tektronix analyzer first and show you how the calculations operate. You're dealing with video signals at 4.2 MHz, and we're also plugging in a 300-kHz filter (Fig. 5-9). Therefore:

$$\text{C/N correction factor} = 10 \log 4.2 \times 10^6 / 0.3 \times 10^6$$
$$\text{(AM and FM)} = 11.46 \text{ dB}$$

Just to be completely sticky, add another 2.5 dB for amplifier and detector losses. The total subtractive factor, then would be:

$$\text{S/N} = \text{C/N} - 14 \text{ dB}$$

Instead of reading a 50-dB video carrier as C/N apparent from tip

Fig. 5-9 *Video signals at 4.2 MHz with a 300-kHz filter on the 7L12 Tektronix analyzer.*

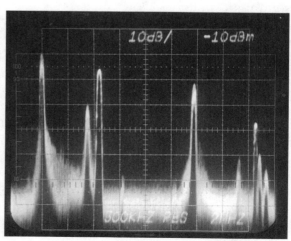

to noise center, the C/N would actually become:

$$C/N_{\substack{terrestrial \\ broadcast}} = 50 \text{ dB} - 14 \text{ dB}$$

$$= 36 \text{ dB}$$

and:

$$S/N = C/N_{true} + 5 \text{ dB}$$
$$= 41 \text{ dB}$$

This signal level is – 20 dBm, which is excellent.

Now we'll do an actual (no fudging) comparison between our 7L12 uncompensated and a true C/N satellite transponder acquisition directly off the air. Take the second transponder from the left, for instance and see how this one reads. We know, for instance, it's a pretty good example, because the TV picture itself is high-quality. We estimate that the number 2 transponder has an apparent C/N of 28 dB, with a signal level of – 46 dBm, which, by the way, is a few dB lower than we'd like. So:

$$C/N_{true} = 25 \text{ dB} - 14 \text{ dB}$$
$$= 11 \text{ dB}$$

Look at the automatic C/N ratio read out by the Tektronix 2710 (Fig. 5-10) using the same transponder, but normalized to 4.2 MHz. Without manual correction or guesswork, the answer appears as 8 dB. However, without a filter, it would have been exactly 11 dB! Presumably, the 2710 is on the nose, but there's only a 3 dB difference anyway. So, S/N = C/N with compensation + 37.5 dB = either 48.5 dB or 45 dB, whichever readout you might prefer.

In the 2710 satellite photo, notice how the 5-MHz bandwidth displays the transponders so much better than the 7L12, which has only a 3-MHz resolution bandwidth. If you doing a 5-MHz resolution bandwidth for satellite work, the difference between the two is obvious. You can't see the modulation very well with only 3-MHz RBW.

As we do more of these TV and satellite studies, the C/N measurement is certainly important, because noise has always been one of the greatest problems in communications. Especially in working with satellites, the dBm power signal level becomes one of utmost importance, because signals at C band must approach – 40 dBm and Ku band at – 45 dBm for suitable video reproduction. Voice and data analog communications might have

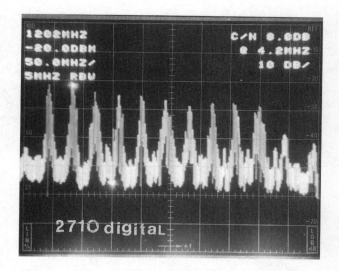

Fig. 5-10 *An eyeball-estimated 28-dB
C/N uncompensated. The
actual C/N is 11 dB (without
video filter). Analyzers are
2710 and 7L12.*

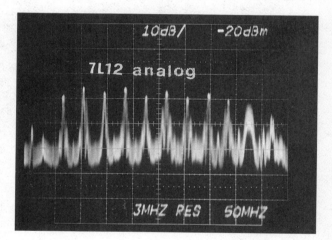

lower power levels as a result of lower frequency content (especially data), because digital transmissions are considerably less noisy than even speech and music transmissions.

In the television RF photo, observe that the voice carrier on the right of the initial video carrier on the left is about 10 dB down and separated by over 4 MHz, and the "suppressed" 3.58-MHz color subcarrier (next to audio) is at least 25 dB down from sound. So when you're working video transmissions, look first for power levels, then pay attention to carrier-to-noise. Observers will readily tolerate a little noise, but they're really unhappy without a picture.

Probably, equivalent "rules of thumb" are available for music/voice/data electromagnetic radiations, which might emanate from the "eye of the beholder" or, more likely, an equipment manufacturer. Regardless, all space/terrestrial transmissions must deliver a reasonable signal level to be adequately detected and reproduced. DBW and dBm power can't be overstressed, for you cannot operate on the ground and in space without them! For example, let's consider a satellite footprint of 38 dBW. Such footprints are either calculated or are actually measured from various ground points and then are issued as signal-strength contours (Fig. 5-11). For our immediate purposes, we'll use new GTE spacecraft GSTAR 4 as our model and work from there.

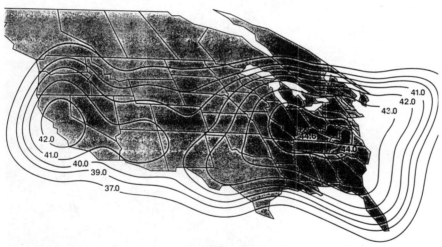

Fig. 5-11 *The GSTAR 4 footprint over most of the continental U.S.* GTE Spacenet

GSTAR 4 (Fig. 5-12) is a 20-watt/transponder spacecraft in geosynchronous orbit with the earth at "parking" location of 1 250 WL. Space loss at Ku is 206 dB, footprint estimated is + 38 dB, antenna gain (at Ku) is 45 dB, block downconverter gains are 55 dB, cable loss (per 100 ft.) is – 9 dB. Add an additional – 7 dB for receiver threshold loss and at least a – 4-dB loss in any heavy rain or snow storm. So, with 38 dBW extended to 68 dBm by adding 30 dB for the conversion, let's see how all this calculates:

$$
\begin{aligned}
\textit{Receiver signal input} &= -206 \text{ dB} + 68 \text{ dB} + 45 \text{ dB} + 55 \text{ dB} \\
&\quad -9 \text{ dB} - 7 \text{ dB} - 4 \text{ dB} \\
&= -226 \text{ dB} + 168 \\
&= -58 \text{ dBm}
\end{aligned}
$$

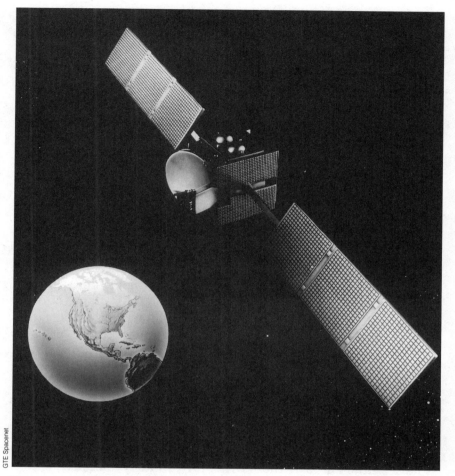

GTE Spacenet

Fig. 5-12 *A GSTAR spacecraft in full flight and orbit, 22.3 kilomiles above the earth.*

Now you have something of a problem. Better receiver/positioners have an effective input signal window from approximately − 60 to − 25 dB; only a 2-dB leeway, even with all parameters included, is asking for trouble.

Block downconverter gains are just about maximum, so no help is there. Antenna gain could be improved slightly, a − 4-dB loss in clear weather would be nullified, and better cable quality could add another + 2 dB. So, we should add at least another + 8 dB under ordinary circumstances, and that would put us in the − 50-dB category, which isn't bad at all. Further, if the receiver threshold could be improved to − 6 dB, and the cable input length reduced somewhat, another 3 dB would be virtually all you could ask for to take care of any inclement weather signal loss.

Of course, all the above is only true for satellite earth terminals with matched feeds and block downconverters, antennas with excellent curvature accuracies, and kingposts that are precisely perpendicular (not even 0.5° out of perpendicular with the earth, that is). Then, you should have a horizon-to-horizon mount with much less than 1/4° Ku access and a positioner that will accept double-read switch counts to help gain this accuracy and a feedhorn that exactly illuminates the entire antenna. If all these conditions are met, you're in like Flynn and possess an excellent Ku-band antenna system.

Although we won't go through the dB arithmetic for TV terrestrial broadcasts, much the same is true for both VHF and UHF, wherever you are. If a spectrum analyzer isn't handy—although you should have one—use a field-strength meter and find out how close you are to – 40-dBm signal levels. With plenty of noise (snow) in the picture, any time your field-strength meter goes below – 20 dBmV, which, translated, turns out to be (at 75Ω) over – 68 dBm, it's a disaster. Matter of fact, field-strength meters—at least mine—don't register below – 40 dBmV, because they were really made for cable measurements, not tiny signals off the air. Some might not applaud that finding, but spectrum analyzers have both sight and scale, and the better ones even have dBmV and dBμV, so little argument should be made. When measuring power levels, trust your dBm readings first, then play with the esoterics afterwards if the spirit moves. Our preference is 50Ω dBm down to baseband, then work with 75Ω to your heart's content. By the way, when matching 75Ω to 50Ω, always subtract 5.72 dB because the transition is a power loss, although you'll only notice a couple of dB change on the graticule scale—a very good point to remember. When you're connected to a signal splitter off the block downconverter or to a TV antenna amplifier, go ahead and make the connection without impedance matching, you're only out a couple of dB anyway.

In completing this part of the chapter, many "learned" dissertations on signal-to-noise are available, but what we've just discussed should satisfy most needs. In some situations, the analyzer's own noise floor conflicts with that of the signal, and you will have to discover it by alternately removing the signal, reconnecting and watching the analyzer, then compensating, if necessary. Usually, this conflict doesn't arise, or we would devote additional space to its theoretical aspects.

Baseband analysis

Having investigated most of the prime RF signals (especially signal levels, carrier amplitudes, and carrier-to-noise) in general terms, it's time to devote attention to audio and video frequencies between 30 Hz and 4 or more MHz. HDTV, of course, is expected to expand the video spectrum to at least 6 MHz (possibly more) following compression and expansion of some 30 MHz companded before and after transmission.

This baseband search will be done with the Tektronix 7L5 analyzer primarily and the 2710, because the 2710 will respond to information down to 10 kHz. Let's tackle audio first, then see what would be of major interest in video. Fortunately, both are AM and FM, including stereo video (called *BTSC-dbx*) that permit the characterization of some interesting signals, rather than stereotypes from antiquity.

Regardless, we'll see where the search leads and report all results with calibrated accuracy; your instrument readouts or displays are only as good as its calibration—a major point for all worthwhile engineers, technicians, and educational laboratory instructors to faithfully remember! A few dB one way or the other often determines the difference between success and failure. However, good electronic design should generally avoid tight I/O (in and out) tolerances if at all possible, because signal processing often adds to thermionic noise (especially with heat).

Noise

Noise quality in electrical devices is customarily measured in terms of *noise figure (NF)*, which can be translated into noise power by the usual 10 log NF. NF, of course, represents the noise figure. In terms of signal-to-noise ratios (S/N), then:

$$NF = \frac{S/N_{input}}{S/N_{output}}$$

Your noise figure is a ratio, which, in terms of decibels, becomes:

$$NP_{noise\ power} = 10 \log NF \text{ in dBm}$$

In dealing with S/N, where the carrier has been stripped and only baseband remains, there will be no equivalent C/N corrections if analyzer noise is below that of the measured signal, and eyeball or cursor readouts are straightforward. So, let's try one or two and look at a little harmonic distortion and power along the way.

First, however, the 7L5 must be horizontally and vertically calibrated by hand so that the readings are accurate. The reference voltage from the built-in calibrator adjusts to – 40 dBm with a center frequency 500 kHz, along with 2 kHz per graticule division, and resolution bandwidth 3 kHz at the – 6-dB down point. Once the L3 input/amplifier is set for dBV, the dBm ranges for 50- and 600-Ω terminators are calibrated also. Therefore, calibrated results appear in Fig. 5-13 at 10 dB and 2 dB/div., exactly as required. Now, we can proceed with a simple sine-wave analysis and see what's possible with the basics.

An output from a rather ''dirty'' sine-wave function generator appears with obvious problems in Fig. 5-14. Notice that the full range of the fundamental (1) is set at – 24 dBm to keep from overloading the mixer/amplifier. If your analyzer doesn't have this window, go to – 20 dBm, but there might be compression. You must also remember that all the apparent harmonics, which actually lessen the spread between the fundamental and harmonic number 2, produce that much more total harmonic distortion (THD). Here the added harmonic levels actually produce a + 15-dB addition to the existing – 36 dB. Therefore, only a – 21-dB difference exists between the harmonic sums and the fundamental. From Table 5-1, you read an 8.9% THD, which is rather poor to say the least. Encouraging, however, is the 66-dB differ-

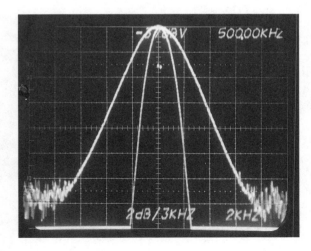

Fig. 5-13 *Although the 7L5 analyzer has digital storage, it is manually calibrated for 10 dB and 2 dB/division.*

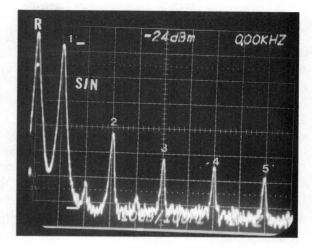

Fig. 5-14 *Frequency, amplitude, S/N, and harmonic distortion (THD) are combined in a single measurement of a "dirty" sine wave.*

ence between the fundamental and the center of noise, which (with no correction) is the S/N apparent, presuming (we hope correctly) that this is the signal noise floor and not that of the analyzer. It's difficult to see, but the resolution bandwidth here is 100 Hz/div. and the per-division setting is 1 kHz. This latter has already been indicated by the position of the fundamental and its reference, and also by precise separation of the various harmonics.

If you only had a 2nd harmonic and no others, there naturally would be no further addition. However, an extra 15 dB of distortion increases the percentage from 2% THD to 8.9% THD. That's almost a major disaster—one that you'd certainly hear in

Table 5-1   dB ratios of THD distortion readings in terms of percentage.

Ratio in dB	% of reading	Ratio in dB	% of reading	dB difference	Add to higher level
20. (40:60)	10% (1%.1%)	30 (50.70)	3.16% (.31, .031%)	Same (OdB)	3.01
21	8.9	31	2.87	1dB	2.54
22	7.94	32	2.51	2	2.13
23	7.08	33	2.24	3	1.76
24	6.31	34	2.00	4	1.46
25	5.62	35	1.78	5	1.19
26	5.01	36	1.59	6	.97
27	4.47	37	1.41	7	.79
28	3.98	38	1.26	8	.64
29	3.55	39	1.12	9	.51
	2nd Harmonic distortions			and Harmonic additions	

Tektronix, Inc.

either tones or multitone music. Sort of a cruddy generator, or what?

Let's change the measurement scale so that the 2nd harmonic and its fundamental rise and fall together, but at maximum amplitude using the same or slightly greater output from the signal generator. Be careful, though, for if the 2nd harmonic begins to rise faster than the fundamental, your analyzer front end will saturate and your readout won't be anywhere near precise. When that point is reached, you have found the apparent power reading, which (in the photo of Fig. 5-15) is $-4 + -52 = -56$ dBm. That's the power output from this form of measurement.

Notice, however, the oscilloscope comparison sine wave—also on the photo above the spectrum information. At 50 mV/div., the scope's trace amplitude is 60 mV. And dBmV = 20 log 60 mV, or -35.56 dBmV.

The analyzer's readout is in dBm, therefore we need to find a dBm equivalent to dBmV, or $dBm = dBV + 30$ (for $dBmV$) $- 10$ log Z (for 75Ω). Unfortunately, this scope's sine wave (even driven to 220 mV) didn't begin to clip until -30 dBm, so the entire project was hopeless because of spectrum analyzer overload.

We probably shouldn't have raised the subject, but you should know that analyzer front ends are extremely sensitive and you must begin your measurements near a level of -30 dBm and not $+16$ dBm. Some analyzers only have front-end tolerances of $+20$ dBm to begin with and this power is but 100 mW. In our example, the scope is only looking at -54.41 dBm—far from any kind of waveform saturation, believe me. Sorry, but it really was

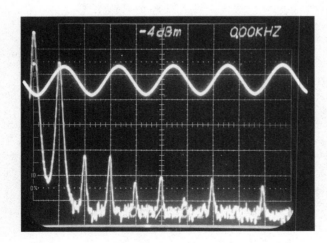

Fig. 5-15 *A clipped scope sine wave isn't an accurate way to measure power. The analyzer's 2nd harmonic is much more reliable.*

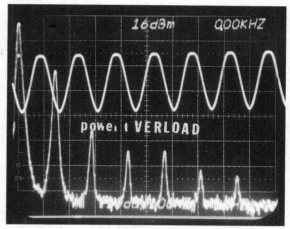

Fig. 5-16 *Forcing a sine wave into compression could ruin the front end of your analyzer; 20 to 30 dBm is the limit.*

necessary to know, so don't use the comparison hereafter. This one, at least, didn't work because of the analyzer's tender front end.

Just for our collective amazement, nonetheless, is the ill-fated power measurement photo (Fig. 5-16) that could have blown the front end of an expensive spectrum analyzer. By cranking up the sine waves, we did (in truth) force the 2nd harmonic to rise faster than the fundamental so that saturation actually showed (this time in dBm) on top of the sine wave. Subtract 46 from 16 and you'll see that this is exactly − 30 dBm. This is why you don't drive your analyzer over − 30 dBm! Some of the purists might want you to hang the fundamental at − 30 dBm on top of the graticule, but then no room would be left to accurately observe any increases, except those of the 2nd harmonic. The 2nd harmonic and the fundamental must reach the top of their ranges without one changing rates faster than the other. One grudging admission: if indicated power approaches watts, you'll have to use a power meter!

A table (Table 5-1) that was published a number of times in my previous books details the correction factors involved. Don't just add the 6-dB difference to 34 dB or you'll be considerably off the actual result. You might also observe that indicated noise is only about 3 to 4 dB and is no factor worth considering.

Now, let's return to that same display and do a power study, if nothing more than to prove that a power meter isn't necessarily your bosom friend when there's an analyzer around. In Fig. 5-16 we've changed the measurement scale from dBV to dBm and dropped amplitudes a few dB because of the more stringent termination. An oscilloscope has also been connected to a splitter so that the analyzer and scope inputs are equal. The trick here is that when generator output causes the harmonics to rise faster in amplitude than the fundamental, that's your maximum undistorted power output. This output is verified by slight rounding (clipping) of the scope waveform superimposed on the analyzer's dBm and 1-kHz/div. frequency picture. Referring to Fig. 5-14, observe that the fundamental amplitude is still less than it was at dBV, but the 2nd harmonic is obviously greater.

As you can see, the analyzer's top reference is 16 dBm, and 46 dB at 10 dB/div., minus 16 dB, becomes a – 30 dBm. The scope's vertical reading is at 0.1 V/div., so it amounts to 0.22 V/div., or 220 mV. Now, because we're dealing in power, millivolts becomes a reference and the scope's waveform amplitude must be reduced to dB(m). So:

$$10 \log 220 = -23.42 \text{ dB(m)}$$

And, compared with – 30 dBm, there's slightly more than 6-dB difference, but both measurements are in the ballpark. With this scope-analyzer approach, however, you don't know quite when to quit with a generator because clipping can occur in different degrees and it isn't a precise means of measuring power. Therefore, stick to the analyzer and you're considerably more accurate. The same applies to amplifier measurements, too. Basics are great, but accuracy is better! By the way, for those still working in dBmV as a power measurement and would like a usable conversion to dBmW:

$$0 \text{ dBmV} = -47 \text{ dBmW} \quad \text{or, as we used to say, just dBm.}$$

Your author isn't exactly excited about calling measurements in volts anything approaching power, because $P = E \times I$, I^2R, or E^2/R in dc and one reduces ac to dc by dividing by 1.414 for peak and 2.828 for p-p. Current does flow, y'know.

Now, let's continue with a few other interesting analyzer applications to fill out the chapter. You can do so much with these highly useful analyzers if you just know how, so let's see where we go from here.

Looking at frequency response

In baseband audio, the easiest way to look at frequency response involves a microphone, a sound amplifier, and a speaker. Simply connect a sine-wave generator to the auxiliary input of an amplifier, advance the volume to a suitable level for the microphone, then connect the microphone to the spectrum analyzer. After that, it's only a matter of beginning with the lowest or highest frequency and working up or down, depending on the selected starting point. Just be sure that the microphone has a fairly linear range and that the amplifier's pretty good, too. As illustrated in Fig. 5-17, the analyzer peak detects and stores simultaneously; you should have a straight line (or nearly so) if all components are reasonably linear. This is also a good way to check your own hearing from 30 Hz to 15 kHz. We continue to use the 7L5 throughout this demonstration, because it is both accurate and linear at the lower frequencies.

This method works, by the way, on both AM and FM receivers because we're measuring the detected signal and not the carrier envelope. If you're not interested in speaker performance, speaker connect terminals or an auxiliary output might be available. At any rate, try the audio spectrum both ways and evaluate all results.

For curiosity, let's illustrate what occurs on an FM stereo radio, then on an MTS (multichannel sound) television receiver. Naturally, we'll have to use different generators because of the TV tuner, but at least the speakers will be the same. Thereafter, the same TV stereo generator will provide adequate signal for a signal-to-noise measurement, which might be quite interesting. Just be sure your own generator's Hz – kHz signals are linear or the analyzer's readout will be false also. This is easily done by running these frequencies across an oscilloscope's graticule to check for any loss or increase in gain.

So you think you have a great speaker system? Crank up the signal generator and find out. Figure 5-17 shows a rather sorry tale. At 300-Hz resolution bandwidth and 1 kHz/div., the frequency response drops rapidly from – 29 dBV down to – 47 dBV at 9.8 kHz, and we only started at 1 kHz. At this rate, perhaps it's rather useless to continue because the drop of over 20 dB (from 2 kHz to 10 kHz) would certainly continue. So beware from whence cometh your audible sound. Actually this response isn't really rotten, but the test isn't exactly welcome news. Fortunately, the glitch at 4.4 kHz is not part of the program; it just happened.

Now go to stereo pickup from the tuner and see what the baseband output of this receiver is—neglecting the speakers altogether. This time, actual RF will be broadcast at 98 MHz so that the tuner, IFs, and amplifiers can all be evaluated together.

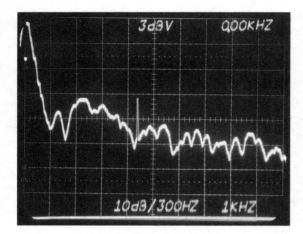

Fig. 5-17 *Wavering speaker reproduction isn't pleasing to the ear, especially 20 dB down at 10 kHz.*

First, however, if this generator isn't expensive and if it can stand a bit of RF calibration, it had best be done before the test or you might never find the audio signal which is, in this case, a 1-kHz modulation. Second, you might just be curious enough to discover if your generator has worthwhile modulation. In Fig. 5-18 at 50 kHz/div., this modulation with a 19-kHz stereo pilot kicks sidebands at 98 kHz center frequency to approximately 190 kHz—plenty for consumer purposes. In addition, if you "kill" the pilot tone, you'll reduce the passband, but show true signal bandwidth with much sharper skirts. So, with a signal level of −40 dBm (remember what we said about TV and satellite receiver amplitudes being −40 dBm for a good signal?) and plenty of passband, your stereo receiver and spectrum analyzer (with whip antenna attached) are subject to adequate signals. In the lower oscilloscope photo, with the scope connected to the composite FM output, the signal with some accompanying noise appears as it does in the lower part of Fig. 5-18. Modulation, by the way, is riding on top of the string of pulses. Also, a speaker outputs a 1-kHz tone that sounds pretty good.

If you check the receiver internally, the composite output can be injected into the receiver's discriminator. Channel isolation is easily examined as we've done before with a dual-trace oscilloscope connected to right and left channels and a $20 \times \log_{10}$

$$\frac{E_L}{E_R}$$

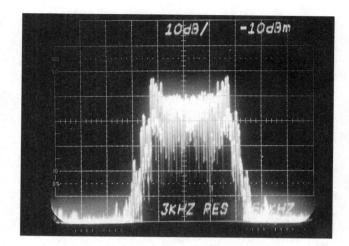

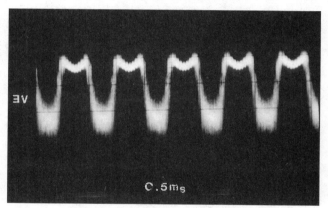

Fig. 5-18 *Simultaneous spectrum analyzer and oscilloscope readouts at RF and baseband.*

ratio of the result. Tone checks and amplitudes are also available at the speaker output terminals. Two or more tones can signify audio bandpass (Fig. 5-19).

You might also like to know that satellite FM signals are virtually identical to the FM display shown (in Fig. 5-18), but obviously they have considerably wider bandwidths as a result of transponder passbands of from 36 to 54 (or more) MHz. One wishes all were as clean as the figure shows.

Lastly, be reminded that FM stereo can be somewhat tricky and both setup and applications must be "on the nose" for good results. Believe it or not, tuning around 100 MHz was a problem until volume was advanced sufficiently to hear the 1-kHz detected tone in opposing speakers, but if you use both an oscilloscope and a spectrum analyzer to do your work, one comple-

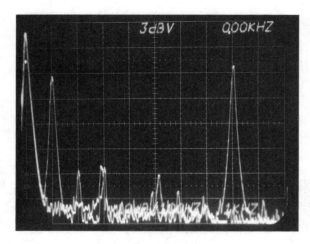

Fig. 5-19 *Two-kHz voltage excursions can reveal stereo-response levels over the range, but not the linearities in between.*

ments the other and results are quite accurate when accompanied by an electronic counter for maximum frequency measurements.

As with any other variety of electrical investigation, results are only as accurate as the operator and his or her instrument calibration. If either are untrained or unattended (serviced), errors will certainly multiply until the final readouts are worthless. This is especially true when tests are remotely programmed for unattended operation. One software slip and the entire investigation is jeopardized.

Multichannel sound

Of course, *multichannel sound* is the name for TV stereo audio, which is derived from the BTSC-dBx system offered by Zenith Electronics Corp. and approved by the Federal Communications Commission on March 29, 1984. Now, however, instead of ordinary frequency modulation for the stereo portion, we have both L + R (left and right channels) and L – R which is double-sideband, suppressed-carrier transmission, turned on by a 15 734-Hz pilot, which is doubled in frequency for the AM carrier. The L + R composite information is processed normally from the composite input, but L – R requires a phase-locked loop detector, a VCO oscillator, dividers, and a special decoder. Together, they produce left and right stereo outputs for further amplification of appropriate speaker systems. Also, a separately detected second audio (SAP) output is normally reserved for FM processing (primarily speech), which has a maximum frequency of 10 kHz.

The second part of multichannel sound (MTS), originates from a dBx encoder/decoder—a noise reduction IC with a possible THD of as low as 0.007%. A dynamic range compressor and expander (compander), it is usually controlled by the rms (root-mean-square) level sensor, which reduces the dynamic-signal incoming range by 50% and doubles its value in decoding.

Presumably, the foregoing remains the basic BTSC-dbx explanation. Probably, a number of circuits have improved over the original as well as, perhaps, in noise reduction and companding has been further simplified, because we note that the Dolby system is indicated.

With the addition of digital processing (such as SEQ and Surround Sound), the audio portion of many TV receivers is considerably more complex electronically, and far superior melodically to sound sections of the '80s. In Fig. 5-20, we have taken actual schematic and block diagram combinations from one of Zenith's new digital receivers to illustrate that which is available today. With PC boards 9-700 and 9-701, we should be able to offer a reasonable explanation of these advanced developments, although there is even more to the audio system.

In 9-701, TV channel information enters the board through J1201, and proceeds to IC 201 via Q1202 and the surface-wave acoustical SAW filter. Here, IF amplifiers provide gain for both video and audio information, detect both, develop automatic gain control (AGC), automatic fine tuning (AFT), and deliver baseband audio to IC1404 and IC1405, which are the digitizing audio ICs.

These two are actually part of the ITT Intermetal Semiconductor Digit 2000 program, which was developed in West Germany and used extensively by Zenith in its excellent sound channels. Identified as the ADC 2300 U and the APU 2400 U, the two were developed specifically for our U.S. standard. The ADC 2300 contains three analog switches for local audio and off-the-air input sources, two sigma-delta modulators, an L-R demodulator for detection of the AM stereo difference signal, and a SAP demodulator for the low-frequency FM information. An IM bus links the ADC 2300 U to its central control.

The PDM I and PDM II sigma-delta modulators furnish pulse trains, whose density is proportional to that of the analog input at a maximum sampling rate of 7.1 MHz. Fully digitized intelligence continues on to the audio processor, which appears on the schematic as IC1405.

The APU 2400 U audio processor converts pulse density into parallel data at the sampling rate via decimation filters. Then,

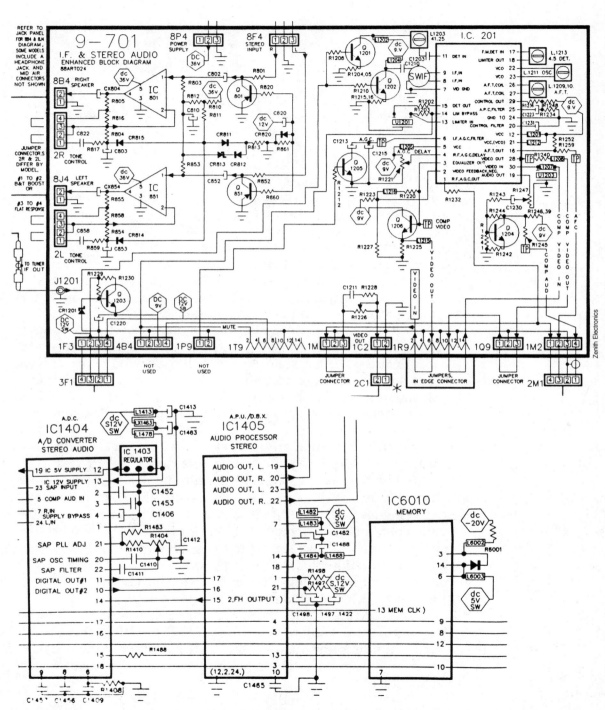

Fig. 5-20 *Partial schematics of Zenith's deluxe sound processing in a consumer TV receiver.*

gain, sum, phase equalization, and deemphasis for L + R are accompanied by L − R into the dematrix mono-stereo section. At this point, L and R convert to pulsewidth-modulated outputs, as well as passing through stereo treble, bass, and balance circuits, which are controlled through the IM bus interface. Another pair of pulse-width modulators (PWM 1) then regulate listening preferences at their outputs.

Zenith did modify the APU 2400 U somewhat to accommodate SEq (known as *spatial equalization*) which widens stereo by cross-feeding L and R channels out of phase, while boosting the bass for flatter frequency response.

In addition to SEq and dBx, top-of-the-line Zenith models also offer Dolby Delta Surround in a passive matrix. First, signals are subtracted for a difference effect, then they are low-passed through a time delay of 10 to 30 ms for ambience, followed by additional noise reduction and output to the volume control.

Because this book is primarily devoted to test and signal instrumentation, we won't proceed further with this audio explanation, but digital sound has become so important to good listening that we thought a general explanation might be of more than passing interest.

Spectrally, the L and R channel outputs appear double-photographed in Fig. 5-21 on the graticule of a 7L12 Tektronix spectrum analyzer, with signal input from a BTSC-dbx stereo generator and appropriate modulation. If the analyzer is programmed to save when these two frequencies are injected via channel 3, the product becomes apparent with excitations at 1 and 8 kHz. Reference voltage with center dot (0 kHz) rests on the extreme left of the graticule. Following the 1-kHz level of 39 dBm, several very small harmonics and spurs pop up, but they're almost 40 dB removed from either signal and will cause no problems.

This information, by the way, can be applied to analog information at any frequency, provided that the analyzer has adequate low-frequency ranges. It's also a very good method to estimate the probable flatness of the spectrum source by mentally tieing the tips of the two pulses together. If you do, then it's possible (Fig. 5-19) to approximate the 0-to-10-kHz response of the audio portion of the receiver with very little effort other than physically connecting the stereo generator and the spectrum analyzer. A two-button push on the generator does this neatly as the analyzer sweeps. Higher frequencies and faster sweep rates, of course,

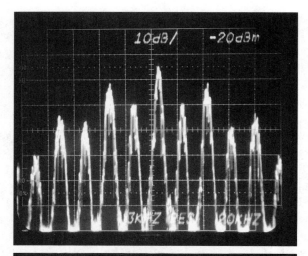

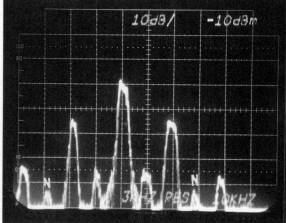

Fig. 5-21 *FM sideband spacings and a 3rd Bessel null identifies modulation that is visible in both photos.*

don't have to be limited to 8 kHz, but with this generator, an external signal is required for additional range. Also, this method is an auxiliary way to check your hearing response and that of the receiver's built-in speakers.

That about wraps up what we wanted to offer in the way of sound for primarily consumer applications, although many more must surely exist in unusual circumstances. Other analyzer considerations, however, bear demonstrating, so let's move on to additional spectrums.

The high end

According to Morris Engleson of Tektronix, three dynamic range specifications are used for most or all spectrum analyzers. *Dis-*

play dynamic range is evident on the vertical graticule. The *1-dB compression point* relates gain to several other dynamic range characteristics. In the third specification, *intermodulation distortion*—functions of a signal input involve third-order intercepts and intermodulation rejection.

These specifications, of course, are primarily design considerations for those with specific requirements and are not of extraordinary general interest. Therefore, we shan't elaborate, but you should be aware of all three considerations because they most assuredly do exist. Mr. Engleson does point out, however, that dynamic range is not related to any analyzer's internal measurements and nonlinearities can permit distortions that the dynamic range does not exhibit. So, beware of exaggerated claims by unknowns. Even though a full-screen range will indicate necessary initial deflection, an analyzer that doesn't deliver this important parameter completely is almost sure to contain problems. Accurate log and linear signal generators with recent calibrations could easily verify measurement difficulties if available. The 1-dB compression point is important, however, because this is an input power limitation and accurate display of both large and small inputs simultaneously becomes extremely important in multiple evaluations of mixed information. Additionally, Tektronix can supply a relationship between dynamic range and intermodulation distortion, if you'd like to calculate the combination in addition to a useful equation for dynamic range, thanks to Mr. Engleson.

Modulation

Meanwhile, instrument accuracies and utilities are dramatically improving as a result of the applied use of microprocessors, extensive software programming, synthesis, special controllers, and a general-purpose interface bus. Now, automatic measurements of total harmonic distortion (THD), amplitude modulation, impulse bandwidth, C/N, peak and quasipeak detect, rms, digital averaging, and so forth, are becoming considerably more commonplace in the better testers.

Percentage of AM modulation is another parameter that derives from the simple equation:

$$\%M = 2Sa/Ca \times 100$$

S represents the sidebands and C represents the carrier relative to the amplitudes (a) of each. When done automatically, this measurement relieves all operator confusion and enables rapid recording. Further, output jacks for X-Y and dot-matrix recorders are extremely valuable for most all stored-type waveforms—especially because they save considerable cash outlay compared to film-associated slides and prints.

Distortions, themselves, are always critical in any measurement process and it's worthwhile to be aware of their cause and effects. The two types of distortions are *linear* and *nonlinear*. The linears usually affect amplitude, phase, and often frequency. The nonlinears present time and frequency difficulties, along with added noise.

A pure sine-wave signal seems to aid nonlinear testing, because it can pass through many networks and allow spurious and allied impure results to appear at outputs without noticeable linear distortion. It's also a very handy method to determine total harmonic distortion (THD) without difficult or unusual equipment preparations. However, THD distortions that approach 0.002% to 0.003% are said to be possible with the better spectrum analyzers, limited basically by the step resolution of their A/D converters. Readouts should appear as true rms values, rather than those identified as average.

Now, because linears add no additional waveform components, the amplitude difficulties and apparent phase shifts should be fairly easy to detect. For permanent record and detailed examinations, stored printouts into recorders would seem preferable to passing CRT displays—especially if the images (or whatever) can't be immediately decoded.

Before reaching for the "wild blue yonder" shortly, you should have just a little additional information on frequency modulation effects, which classically demonstrate the Bessel functions described earlier. FM deviation normally has to be calculated, based on the Bessel null method of measurement. Here, the total power remains constant, but not the deviation.

By using the equation that can produce t for the modulation index, delta F for deviation, and f for the modulating frequency:

$$\text{delta } F = tf, \text{ or } f = \text{delta } F/t \text{ and } t = \text{delta} F/f$$

Then, return to the Bessel null table and pick your number selection. However, if you'd like all this done without pain or strain, select a good spectrum analyzer, program it for whatever null you'd like, and the job is accomplished automatically.

The illustrated example in Fig. 5-21 shows that sideband

spacing equals the modulating frequency, but nulling the carrier allows deviation calculations at whatever Bessel function is selected. The initial portion of the figure exhibits a 100-MHz carrier with the 19-kHz pilot turned on and 1-kHz internal modulation into L − R. Also two pairs of sidebands are evident and, as you should remember, increase their numbers according to the 75-kHz peak deviation of the carrier, divided by the highest modulation frequency. Or, modulation index = carrier peak deviation/maximum modulating frequency. The nulling frequency can now be determined by adding the external input of an audio generator in parallel with a frequency counter and tuning for Bessel nulls (given in the modulation index table). When a modulation frequency nulls one of these sidebands, you have both the exact frequency readout on the analyzer and also the modulation index number (Fig. 5-21). So, go through this exercise and see how the numbers correspond. Notice that you must begin with at least the 2nd null because 75/2.404 8 equals 31.19 kHz, and you're only allowed a 15-kHz maximum frequency response in FM deviation.

We have no guarantees that our slightly suspect, low-cost signal generator is accurate, but the recorded frequency that produced a touchy beat was 9.493 kHz at the third null. A strict Bessel calculation would say it was 8.653 1. So, the answer differs by almost one kilohertz, but let's continue:

$$\text{The modulation index, } M = 75 \text{ kHz}/9.493 \text{ kHz}$$
$$= 7.90$$

You can play, substituting various numbers into the initial equations above or simply use the signal generator/counter method, which automatically gives a final result anyway. Suit yourself. However, by using external modulation only and adjusting the generator's level very carefully, the Bessel nulls do appear and the modulation index, after that, is extremely simple—just find the two parameters and solve for the third. Throughout the project, incidentally, the FM carrier remained at 98 MHz, regardless of generator modulation—whether internal or external. By-the-by, if external modulation seems to change nothing, open up your modulating generator and find the trouble—mine was a simple regulator. Unfortunately, those things do happen to even the best of us. The 98-MHz carrier was delivered externally via a monopole antenna.

The foregoing again proves that basic measurements aren't

difficult once you understand the application and possess test gear of reasonable quality. Obviously, if you're demanding 4-place accuracy, instrumentation will have to be costly and fully calibrated, but an approximation is useful and puts you in the ballpark, usually without fail. One considerable admonition to one and all when working with either inexpensive or, especially, older signal generating equipment: check dc voltages to be sure that they correspond with the manufacturer's values, take the first one or two steps recommended for calibration, then use a good electronic counter with at least a 1-MΩ input impedance for the rest.

Many of the old instructions depended on analog voltmeters and 60-Hz references for their odd-ball standards, which small shops and amateurs did actually possess. Today we are either aware or have discovered that neither 60-Hz nor nondigital instruments are exactly reliable; twelve-place counters and 4.5-digit voltmeters are! Next time, we'll calibrate first and measure afterwards; then, the accuracy numbers should correspond.

The 492PGM analyzer

For those of modest means, the 492PGM by Tektronix is indeed, a formidable instrument. Fully synthesized, with a frequency range from 10 kHz to 21 GHz without extenders, the A-B storage and center frequency can be selected by either the Center/Marker knob or by the Data-Entry pushbutton. The single and delta markers deliver accurate delta readouts of frequency and amplitudes, and all front-panel control settings so intended can be remotely supervised via the built-in GPIB port. A number of systems and controllers are thereby matched and used via the GPIB for programmed operations with this analyzer.

To give you an idea of the accuracy of the instrument after a 1-hour warmup, the phase-locked 1st local oscillator has a typical error of less than 50 Hz/minute, and a readout resolution of less than 10% of span/div. The 10-dB log accuracy is basically ± 1 dB, the 2-dB log mode is ± 0.4 dB, and the linear mode is $\pm 5\%$ of full scale. However, with a weight of 47.5 lbs., it can't be treated exactly like the tournament brass ring. A front-panel drawing of the equipment illustrates its qualities in Fig. 5-22. In the digital-storage mode, two complete displays can be digitized and stored. In store display and recall conditions, as many as nine selected information items can be saved and recalled, with an error of only 0.5% full screen.

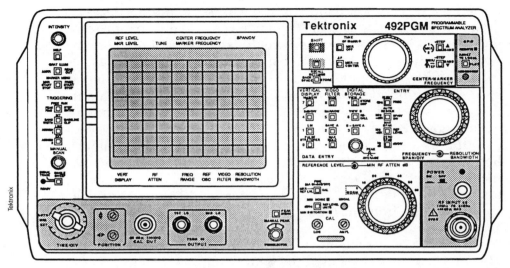

Fig. 5-22 *Front face drawing of Tek's new 21-GHz portable spectrum analyzer with all of the "bells and whistles."*

If you have questions about control commands, operations, or even error detection, a Help pushbutton can be engaged and an on-screen reply describes the particular function. If some portion of the information on-screen requires even closer study, a Manual Scan control is available to do just that. If it has too much analyzer noise, video filters restrict bandwidths and reduce noise—especially when two signals are close to one another. Then, if you have a question of true signal identity versus false signals, special Signal-Identity mode local-oscillator frequencies are shifted on alternate sweeps. True information shifts only slightly, while spurious signals often occupy one or more separate divisions. In other words, a small shift identifies what we're looking for.

As with many other analyzers, triggering can free run, supplied internally by the video signal, at line frequency of the power supply, externally through the EXT-In connector on the rear panel, or as single-sweep (following trigger arming).

If pulse signals require analysis, select a wider bandwidth than that automatically supplied. For non-digital storage information, set the sweep rate for minimum flicker. In Max Span, the full frequency range of the instrument is displayed. Otherwise, the analyzer will normally control the selected sweep and the set resolution bandwidth for a continued calibrated response. The objective of Resolution Bandwidth, of course, identifies discrete frequencies within a certain frequency span, and as bandwidth decreases, S/N ratios increase, which delivers maximum sensitivity via minimized resolution bandwidths.

Normally, the input impedance of the 492PGM is 50 Ω, but a special matching pad or Option 07 can also offer a 75-Ω input (usually used in CATV or video measurements, where impedances are normally of the higher value). Remember, however, that dBm is always a power measurement; dBV or dBmV identifies voltage only or other similar spectra in the 75-Ω category, including audio. If you would attempt voltage measurements under power conditions, only few on-screen dB would separate one impedance from the other, and a matching attenuator should always be used.

Warning: Do not attempt to introduce dc into this instrument or your repair experience will be both painful and significantly expensive. Always have a dc block on the input before signal introduction. You might also observe that sensitivity levels are usually measured in dBm, because this is dB, with reference to 1 mW. Analyzer operational span extends from − 120 to + 30 dBm or 0.001 pW to 1 W. Keep all connecting cables as short as you can to avoid impedance mismatches, degraded sensitivities, and reduced frequency responses. First-class spectrum analyzers are not toys and their performance is usually just as good as their operators!

Action in high places

It's time now to abandon characteristics and superlatives and let you see what this instrument can actually do with the help of some lively signal inputs in MHz and GHz. In this way, we can demonstrate some of the important features considerably better than unexciting descriptions.

First, let's test the calibration of a nonsynthesized signal generator (Fig. 5-23) commonly available in the lower-cost market. At − 20 dBm, the signal-level readout is − 29 dBm, the frequency is 85.55 MHz, frequency-per-division is 500 kHz, the resolution bandwidth is 100 kHz, and the scale factor is 10 dB/div. Most of the noise in this instance is that of the analyzer (connected and disconnected), but the sine-wave CW signal appears very clean at this frequency, with a dial reading of 86 MHz. For a BK-Precision Model 2005 generator that sells for $229 this accuracy certainly isn't unreasonable—especially because a two-hour

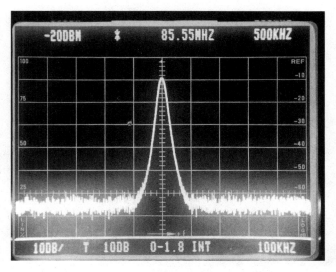

Fig. 5-23 *As frequencies rise toward and beyond 100 MHz, signal and display accuracies become that much more important. RBW at 6 dB down is 100 kHz.*

frequency "soak" only produced a change of 50 kHz. CW signals, as you can see, are extremely important from time, amplitude, and frequency measurements, and also when quantifying special test results (such as those used in evaluating microwave and even consumer parabolic satellite antennas). Gain and side-lobes, especially, with high-frequency analyzers that possess very small resolution bandwidths, can deliver excellent readouts with good accuracy and reasonable utility.

The next step involves signal generators with harmonics. In Fig. 5-24, this generator does quite well at 256.45 MHz—almost an exact 3rd harmonic of the frequency selected in Fig. 5-20 (only off by 200 kHz). CW amplitude, however, has dropped 10 dB, and will probably continue to diminish as we increase toward 500 MHz. Along the way, we also noted that the 2nd harmonic was apparent at about the same amplitude.

Now let's go to 400 MHz (Fig. 5-25) and analyze the effect there. Don't expect maximum results at either end of a low-cost signal generator's range because they aren't there. Although the frequency accuracy of 403.26 MHz versus the 400-MHz signal generator readout isn't bad, look at what happens when the resolution bandwidth drops to 10 kHz. All sorts of spurious transients creep in, the top of the display arbitrarily widens (even though waveform amplitude returns to − 28 dBm), and the noise floor decreases to less than half its former value.

These are just a few of the items worth evaluating in any signal-generating test gear that needs to be investigated before you

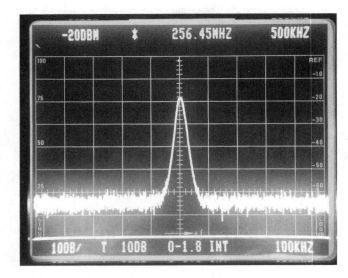

Fig. 5-24 *The third harmonic of 85.55 MHz is almost on the button, except that its amplitude has dropped 10 dB.*

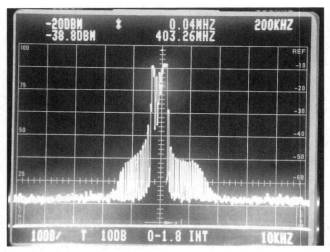

Fig. 5-25 *At 400 MHz, the original amplitude does not change, but waveform distortion is clearly indicated.*

go to the bank. The −38.8 dBm on the left is the instantaneous marker amplitude as the various waveform spurious bobs and weaves continue. Observe, too, that the per-division readout is now 200 kHz, instead of 500 kHz.

Sweep generators are the next topic. Here, with the generator's variable amplitude retarded, the lower trace of Fig. 5-26 looks reasonably linear—certainly within a couple of dB and at an amplitude of approximately 8 dB (but −52.8 dB down). Recall that TV tuner and satellite inputs tolerate no more than 60 dB down (worst case), and that for a reasonably noise-free picture

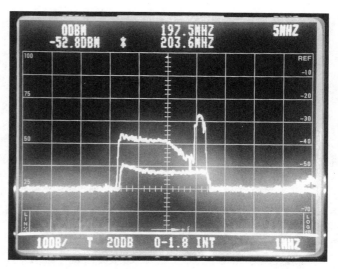

Fig. 5-26 *Sweep generators are obviously not always linear. This one was designed to sweep to 900 MHz. The upper trace was not usable at 200 MHz!*

with relatively good definition, – 40-dBm levels are desirable. Then, see what occurs when the sweeper's amplitude is opened wide. Sure, you pick up another dozen dB at about 190 MHz on the left, but the sweep is totally nonlinear—with a drop of 10 dB on the right and then a huge shaft of nonsignal on the far right (at approximately 204 MHz). This display is not a double exposure, but results from A-B digital storage, which is developed on a single photograph.

So, instead of dual-channel inputs, the 492PGM has adequate dual-trace storage and sufficient maximum hold to accommodate entirely separate voltages, which illustrate two completely different sweeps. Our question is, therefore, would you accept this weak output and nonlinearity for your expensive sweep generator? For this reason, it's best to check first and ask questions afterwards. Just shows what a good spectrum analyzer can do!

This particular generator is supposed to respond all the way up to 900 MHz. Therefore, you might be somewhat more than interested in the following results. With the top frequency read-out being that of center frequency and the lower one that of the marker, you can see that the first has a level of – 62 dBm, which is patently unusable (Fig. 5-27). Note also that even with maximum and minimum amplitudes adjacent and virtually adding, the sweep span is less than 5 MHz; this is allegedly a 15-MHz sweep generator. Lest you suspect the last two figures are phony, let me assure you, they're not! Imagine what your author has undergone over these past 10 or more years, trying to align

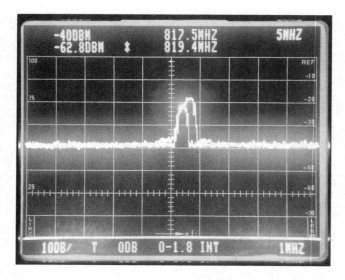

Fig. 5-27 *At 819.4 MHz, this particular "sweep generator" displays little more than a pair of badly distorted CW signals at no more RBW than 2 MHz.*

receiving equipment with such a monstrosity. Friends, it hasn't been easy.

Consequently, when working with any test gear of less-than-professional quality, see if you can't at least borrow a real spectrum analyzer and check it out before you buy one. It saves a lot of headaches and removes many inaccuracies from your rubbery conscience.

What you've seen in the preceding paragraphs aptly describes the many problems with low-cost instrumentation of only a few years past. Today, they've improved considerably, but you still can't "make a silk purse out of a sow's ear." She just wasn't built that way.

"From the ridiculous to the sublime" best describes our introduction to Hewlett-Packard's 10 MHz to 20 GHz Model 83620A sweeper that was borrowed with special dispensation from HP'S Jan Whittacre in the state of Washington and Bob Chase in Maryland. Just look at the beautiful 900 MHz to 1.45 GHz display in Fig. 5-28. Isn't it a beauty? Also notice the perpendicular risetimes and falltimes, where the sweep begins and ends. At a deflection of over 600 MHz, you can certainly look at many high-frequency conditions—especially the satellite 950 to 1 450 MHz C-band channel bandwidths that fall nicely into that familiar category. Although an expensive 20-GHz generator might seem to be deliberate overkill, we'll try a measurement or two that can certainly justify its good services.

Strangely enough, we were able to set up this instrument

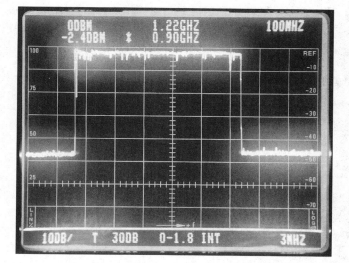

Fig. 5-28 *A beautiful sweep for HP's 83620A as it covers the band from 950 to 1 450 MHz (the satellite block downconverter passband).*

without even a glance at its operator's guide. Just a little different from its lesser kin, time sets, sweep span, frequency determinations, and dB output are all easy to locate. Readouts are then viewed in green letters on a jet black background.

Let's maintain this same sweep time over the 500-MHz satellite range (Fig. 5-29), but make two significant equipment changes: instead of 0-dBm output, we'll adjust the generator for + 20-dBm output and connect directly to a fairly decent C-band feedhorn with no amplification. We'll hook up another feedhorn and its 50-dB gain-block downconverter 25 feet across the room.

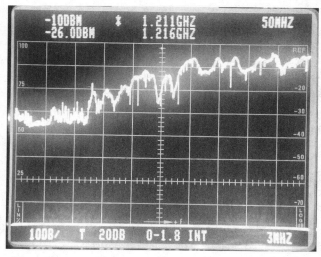

Fig. 5-29 *A satellite feedhorn responds to laboratory stimulus with Tek and HP superior equipment.*

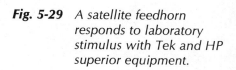

Then, 18 V is applied to a dc block, which passes all signals, but prohibits any power-supply voltage from reaching the Tektronix spectrum analyzer (connected via cable to the downconverter). The analyzer's input now reads – 26 dBm at 50 MHz instead of 100 MHz per division. We can now see the feedhorn and block downconverter response over the full 500-MHz span that is allotted to each C-band satellite.

Observe that a 20-dB difference is between the 900-MHz and the 1 450-MHz portion of the swept display, which indicates much better gain at the high end with a few flips and dips along the way. Obviously, the display is far from linear, and the test method is somewhat suspect. However, you can certainly see that even a somewhat crude laboratory investigation immediately tells many amplifier and feedhorn characteristics within a single satellite parameter.

If you want to do a little arithmetic, cable losses are about – 6 dB added to – 26-dB marker indication, which is – 32 dB. With the generator delivering + 20 dBm to the transmitting feedhorn, that's a difference of 52 dB, which is virtually the gain of the receiving feed's block downconverter and its built-in amplifier. Now you can see that similar evaluations, either simple quickies or careful, precise setups, can be highly useful— even within the confined space of an electronics laboratory.

New Tektronix offering

It's new, it's different, and frequency response and utility are even wider and greater than ever (Fig. 5-30). Furthermore, the membrane panel has been replaced by pushbuttons, and it has more standard features, up to 108 displays, 36 front-panel settings, selected keystrokes are available via a Save-Enable setting, and a special jack for built-in AM/FM detectors, an amplifier, and a speaker headphone jack. The resolution bandwidth extends to 5 MHz, so an Option 10 video monitor can easily be installed for terrestrial, as well as satellite RF demodulation of picture information. Quasi-Peak is optional.

What's its name? The new 2712, portable 9 kHz to 1.8 GHz, of course. As always, it is programmable, just like its forebearer, the extremely popular 2710. The 2712 now features input sensitivity of – 127 dBm or – 139 dBm with a built preamplifier up to 600 MHz, a signal counter that's accurate to $1/2$ ppm, a four-trace display storage, an automatic waveform IEEE-488 bus to printers

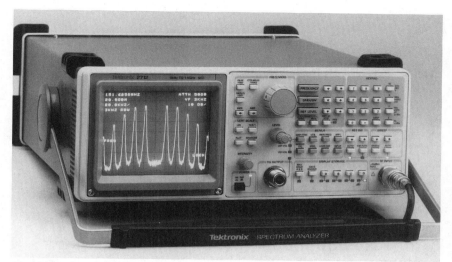

Fig. 5-30 *A mid-1991 offering from Tektronix, the 2712 is a deluxe version of the 2710, but with buttons and many former options included.*

and plotters, and an 80-dB dynamic range with optional battery pack (weight is just 21 lbs. without and 39.5 lbs. with).

In the end, however, it's how this equipment performs in its natural environment, rather than in some protected space that demonstrates its real worth. Regardless, preliminary testing can always be valuable in either design or troubleshooting, and the combination of a superior sweep/signal generator and an excellent spectrum analyzer can make the GHz frequencies seem almost like MHz in any season, and solve a great many problems and perplexing questions for you!

❖ 6
Logic analysis in state and timing logic

WITH THE FURTHER EXPANSION OF VLSI AND AN ANTICIPATED explosion of reconfigurable logic, timing and state analysis instruments are assuming even more significant importance than they have in the past, and further projections of added utility in the future. For example, Philips/Fluke has already announced a simultaneous state and timing analyzer using single probes. This system saves both time and accuracy because removal of the second channel probe "eliminates a source of measurement errors" with capacitance loading of only 8 picofarads maximum.

In addition, setup and applications for these multiprobe instruments are becoming considerably more user-friendly as designers accumulate further field feedback, faster logic ICs, appropriate programmability, and just plain experience. With the influence of analog decreasing and that of digital increasing, including digital TV receivers, high-definition television, and fiberoptic transmissions, logic analysis, ahd troubleshooting become a necessity throughout the communications world.

Now, with static memory cells readily available for programmable logic, "soft" hardware in the form of reconfigurable programs offers designer and certain users the option of actually changing I/Os, switching interconnects, and working with a considerable array of logic operations.

For manufacturers, programmable logic offers added reliability, quicker design-change response, less risk, and attractive field modifications that could save many a system from becoming outdated or fix ongoing problems. Here, self -diagnostics, such as you were initially exposed to in the DSO chapter, would be another good offering, which, however, could actually be reconfigured to test other systems when the need would arise. As you might surmise, many more applications for soft and hardware logic have been and are being discovered as additional systems reach the market. All of this is well worth investigating for anyone in the applied logic business. Its use is bound to increase as hardware and software compete as a result of more advanced engineering and imaginative programming.

Why analyzers?

Gone are the days of *discrete* ANDs/NANDs, OR/NORs, Wired ANDs/ORs, exclusive ANDs/ORs, and implied ORs. A few others are tossed in for a good exercise in *Boolean algebra*, named for George Boole, a mathematics professor at Dublin University during the middle 1850s. His 1s and 0s were the same as they are now, except for logic refinement and progress during the 140-odd years since. Nonetheless, Professor Boole's laws of combination, unique elements, negation, commutation, expansion, absorption, etc., remain just as conceived years before the American Civil War, as do the Karnaugh maps, deMorgan's Theorem, and double negation.

But if the Karnaugh map is glitch-free, so the wired-ANDs and implied-ORs are predicated upon the sum-of-products that have no static hazards produced by between-state switches developing an identical output state. By any other name, a *glitch*!

More recently, we've passed through the initial era of field-programmable logic arrays (FPLAs), asynchronous sequential logic, digital-code converters, computer-aided logic synthesis, dual independent feedback, programmable flip-flops, multi-mode counters, on-chip diagnostics, EPLD (electrically programmable and erasable logic devices)—logic cell arrays that contain configurable logic blocks, I/Os, interconnects, time-shared logic, protocols for system exchanges, programmable electrically erasable devices, folded NANDs, the usual JK and D flip-flops, PLDs and on-board synthesis, microwave logic that operates beyond 2.5 GHz, and even convolutional coding.

All this generates a substantial need for test equipment that can engage systems and subsystems logically in both timing and state analysis. These state analyzers compress information into character or number lists—often binary, ASCII, octal, or hexadecimal (hex).

Often called a "smart multichannel digital recorder," logic analyzers usually display waveform timing, digital words in state, and disassembly for microprocessor operations codes. Horizontal scales are adjustable, data-point time differences can be measured, waveform expansion (zooming) is possible, and state analyzers code timing data into digital words via tubular format in hex and timing information shown vertically. Some analyzers also permit translations into additional formats such as octal, binary, American Standard decimal (ASCII). The latter is a 7-unit digital code for teleprinters or teletyperwriters. Other analyzers can select desirable information to record along with stop and start times—either synchronously or asynchronously, depending on triggering by internal or external clocks. Usually, asynchronous sampling will measure timing patterns; synchronous sampling is useful in digital state configurations. Sampling speeds up to 100 MHz are more or less typical, unless special needs or particularly fast systems require investigation. High-end analyzers can be exceedingly expensive, but they accept many more channels than normal and have sampling speeds into the GHz.

Logic analyzers occupy major roles in debugging hardware, as well as software that operates on microprocessor-based systems. Small ROM systems are easy, because codes are available at known addresses. Time-shared complex systems are considerably more difficult, because data flow must be monitored and analyzed in real time.

Fortunately, the logic analyzer (LA) can actually document data flow and trigger times for either local or manufacturer evaluation. For difficult situations, floppy disks can accumulate information over considerable periods and even telephone connections are available for those who are further sophisticated. Single node and/or complete board testing is possible with such additions, and new products without service records can even be recorded for eventual troubleshooting programs. With later factory backup, many seemingly insoluble programs are often corrected very quickly. In microprocessor-based systems, multiple reception and command means critical timing. This alone can generate a host of related (or even unrelated) problems—especially because either hardware or software can be the culprit.

Timed measurements are largely dependent on internal clock speeds to measure and record time relationships among various signals. For 8-bit microprocessors, for instance, a 10-ns resolution is probably sufficient, but faster units will certainly call for additional speed. To acquire information at the precise time an error appears, the analyzer's trigger mechanism must recognize and store both the problem and the surrounding data, and then stop. Sufficient memory to do this is most important, triggering must be fast, and special glitch detectors might be required. Very narrow glitches can easily necessitate 400-MHz frequencies with a 2.5-ns resolution. As a further example, we're told that a 32-bit microprocessor might well need 72 lines for state information, 24 lines for timing, and up to 4 lines for clocks. So, a 100-channel analyzer isn't too much to expect, considering the equipment now coming off the assembly lines.

Lower-cost logic analyzers (under $5,000) are appearing and they will often accommodate between 32 and 64 inputs. Useful in production tests, R&D, service, schools, and sometimes design, both quality and applications are improving, which allows further expansion throughout digital electronics. In addition, more equipment allows both state and timing data to be acquired at the same time. This equipment means additional accuracy, fewer required channels, less-capacitative loading, and any channel can become a clock, qualifier, and a timing signal. In at least one Philips LA, a separate glitch detector operates in parallel with the timing analyzer, which permits glitch stores as low as 3 ns, with the timing analyzer running wide open in the transitional mode. Also, a movement is afoot to integrate DSOs with logic analyzers, now underway with Tektronix.

Necessity and importance are becoming closely associated with analyzers of many descriptions. Tektronix continues to believe that logic analyzers must be the instruments for recognizing and isolating logic faults and that viewing the actual waveshapes is necessary before making a final diagnosis. However, there are difficulties along the way in integrating the troubleshooting equipment and further progress should become apparent by the time this chapter is published—possibly even before, according to Tektronix.

Easier is often better and less threatening

In the meantime, improved interfaces, microprocessor pods and adapters, and appropriate menus and display CRTs are making analyzers considerably easier to operate—especially when microprocessor codes exceed 50 MHz. Now, RISC (reduced instruction set computer) and CISC processors can be debugged. DSP (digital signal processor) chips have become increasingly useful, especially when equipped with a good disassembler, ROM emulator, and embedded debugger. These features are mostly contained in the higher-priced instruments, so a mention of such special abilities is about as far as we go here because they're primarily used by designers, not troubleshooters.

However, fast setup and special hold times are important for all of us, as are the pitfalls of complex clocking curtailing speeds because special glitches can lower speed, channel count, or possibly state memory, according to John Fluke Manufacturing. Therefore, read the fine print in the specifications before you acquire so that compensated measurements are unnecessary. Nonetheless, the logic analyzer is really the basic instrument for integrating hardware and software, even though the software-design contingent has a built-in fear of hardware. The simpler its setup and use becomes, the more rapidly its acceptance will be for all members of the digital trade.

Performance and state analysis are basic factors in system evaluations. In conjection with digital-storage oscilloscopes, not only the 1s and 0s can be seen, but also the actual waveshapes themselves. Beware, therefore, when considering additional plug-ins for various options that are always offered. Some might support good measurements, but others can detract from the initial goal and actually impede a particular result. Look on an LA as a problem solver, not a special college course in logic's delight. If all else fails, ask the applications engineer for aid and any special literature. Both are usually of inestimable help before and after the sale. If not, go see someone else. Many should be called, but few chosen. Beware!

With those unguarded words of general introduction, it's now time to move on to the actual equipment and see how they prove their worth in specifications and performance.

Hewlett-Packard

Always among the first in sophisticated instrumentation, HP (Fig. 6-1) is often a bit more expensive, but its equipment performs according to specifications and sales engineers are extensively and superbly trained in applications and assistance. Further, frequent class instructions for technicians and engineers

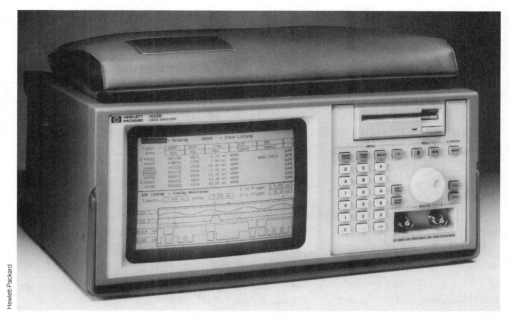

Hewlett-Packard

Fig. 6-1 *HP's 1652B logic analyzer with 80 channels and 100-MHz timing on both channels.*

are sponsored by Hewlett-Packard in many of the larger U.S. cities and conducted by top-flight engineers, who use abundant user-friendly literature and notes. A few minutes of reading time can often be very beneficial, especially when working with application notes and service communications. To use any test gear, you should at least know its better performance ranges and also its limitations. Nothing, fortunately or otherwise, seems to be perfect—at least not in the past 2,000 years.

For instance, if you're planning to test all CMOS and the majority of ECL operations, want time-interval accuracies of better than ±150 ps, and 18 channels at 1 GS/s, then you'll have to consider an HP 16532A oscilloscope module for the HP 16500A logic analyzer at a little under 10 kilobucks. However, if somewhat less-exotic instruments will do, then you might well select one of our picks, the HP 1652B, which also offers a dual-trace 400-MHz digitizing oscilloscope. With this advantage, you can look at details of the waveforms along with timing at 100 MHz for all 80 channels, and state at 35 MHz for the same. This permits analysis of 8-, 10-, and/or 12-bit microprocessors with time intervals of 1-ns accuracy and evaluation of glitches in terms of noise and loading.

With 400 MS/s, both channels are capable of capturing one-shots that have 2 048 samples/channel and precision time-interval measurements, including autoscale, auto-calibrations, plus averaging display, along with time-correlated state, timing, and analog information together in a single portable package that weighs only 24 lbs. Add full triggering, built-in disk drives, and data configuration files that are compatible with the HP 1650B, 1651B, and 1654B so that information can be transferred from one analyzer to another. Of course, it has hard copy (printer) output for more than 10 printers. You will also find automatic pulse measurements for pulse width, frequency, risetime, p-p, preshoot, overshoot, falltime, and period, and the 2K sample memory/channel allows the viewing of waveforms up to 5 μs before the trigger, retaining greater than 1-ns time-interval accuracy.

These features, of course, are extremely useful when measuring any and all pulse parameters—especially when microprocessors appear to be at fault. Buses that serve such processors and connect logic have specific risetimes and falltimes on command and data routes. When processors and drivers age or power supplies slump, communication tracts become unreliable and errors are commonplace.

Some scopes will automatically read out minimum to maximum waveform values, the time between, and also find the rise/fall 10 to 90% points. Delay times between two channels are important, too, and you will discover that automatic pulse measurements are usually more accurate than manual cursor settings, because they're based on digital sampling rather than on a CRT display. Also because DSOs can see pretrigger logic patterns, they can supplement and confirm logic analyzer readouts and allow an actual view of existing voltages (whether precise or otherwise).

If data bit streams become jittery or pattern sizes considerable, the logic analyzer can quickly come to the rescue, and hold-off set times can also be synchronized to the DOS's clock with surprising resolution. Does the above suggest some specific advantages to combining a good, fast DSO with a worthwhile logic analyzer? You bet! And, in time, you can expect multicolored displays to feature the best in both instruments, as further engineering design develops.

Practical aspects

When you have up to five clocks of qualifiers that permit precision sampling, coupled through eight levels with storage abilities, and eight pattern recognizers that permit data extraction from complex codes, you then have a powerful analyzer indeed. Afterwards use built-in disk drives for field data accumulations, then study the results at leisure. During many data checks, you can also program the analyzer from either the HP-IB or RS-232 ports and rely on the spare unit to control the printer. Passive, lightweight probes connect directly to the system under study with no more loading than 100 kΩ shunted by a 8 pF.

In the glitch mode, sample periods are from 20 ns to 50 ms, and sample accuracies are 0.01% of the sample interval. Analyzers, such as the 1652B, can be armed by the run key or by the built-in oscilloscope, and channels can be grouped into a six-character name. We should continue, but oscilloscope specifications have already been lightly examined and operational considerations need to be discussed, too.

Triggering in state analysis

Data from the circuit device (or system) under test (DUT) provides clock triggering via, let's say, pod number 1. To set up in HP's system, press the Format key and adjust the J-Clock to sample on a negative-going transition. Next, the cursor is placed in the Clock Field by pressing Select (Fig. 6-2). The trigger is then "told" to begin operating at address 0000 on power up. The Run key both arms the state analyzer and then forces a reset (Fig. 6-3). State listings now display three vertical columns of alphanumeric numbers that identify hexadecimal logic.

In this listing, the numbers on the left identify line numbers, which might be positive or negative. These values denote which states occur before the trigger (–) and those that occur afterwards (+). The states appear in the second column (marked *ADDR*), and the third column denotes those states that the analyzer sees on the data bus (appropriately labeled *DATA*).

We could continue step-by-step until valid and/or incorrect data are uncovered and verified, but a number of other conditions can be observed with broader information applications to

```
68000STATE - State Listing
Markers      Off
```

Label >	ADDR	DATA
Base >	Hex	Hex
−0007	0088C4	4E75
−0006	0088C6	61E6
−0005	0004F0	0000
−0004	0004F2	88C8
−0003	0088C8	B03C
−0002	0088CA	00FF
−0001	0088CC	6730
+0000	000000	0000
+0001	000002	04FC
+0002	000004	0000
+0003	000006	8048
+0004	008048	2E7C
+0005	00804A	0000
+0006	00804C	04FC
+0007	00804E	61D8
+0008	008050	6100

Reset Vector Fetch Routine

Fig. 6-2 *Setting up the analyzer preparatory to receiving data.* Hewlett-Packard

```
68000STATE - State Listing
Markers      Off
```

Label >	ADDR	DATA
Base >	Hex	Hex
−0007	0088C4	4E75
−0006	0088C6	61E6
−0005	0004F0	0000
−0004	0004F2	88C8
−0003	0088C8	B03C
−0002	0088CA	00FF
−0001	0088CC	6730
+0000	000000	0000
+0001	000002	04FC
+0002	000004	0000
+0003	000006	8048
+0004	008048	2E7C
+0005	00804A	0000
+0006	00804C	04FC
+0007	00804E	61D8
+0008	008050	6100

State Locations

Fig. 6-3 *The state listing of number locations on address and data buses.*
Hewlett-Packard

the general world of logic analyzers. Here, you are subject to a basic trigger command and search in state analysis, just to get the ball rolling. These directions are all contained in the HP 1652B *Getting Started Guide*, in which the company suggests how to become acquainted with this useful analyzer.

State analysis was deliberately selected for our example, because it is by far the more complex of the two analyzer modes. The timing portion is quite simple by comparison. Therefore, master state first, and the remainder becomes second nature if you know anything about an oscilloscope—especially a DSO.

Innovators

Although no one has yet combined a DSO with a logic analyzer that can display all three analytic modes simultaneously on its cathode ray tube, at least one major manufacturer has combined both digital readouts and might well induce the remainder of the industry to quickly follow suit. In the meantime, memories are becoming longer, sampling faster, triggering more readily available where needed, and costs moving downward as better integrated circuits, competition, and prices influence the market. However, as huge numbers of complex ICs of all descriptions, especially digital, are poured into virtually every military, commercial, and consumer electronic item, the absolute need for logic analyzers grows proportionally. He who corners even a reasonable share of such analyzer test gear has acquired some hefty pocket change. Later, as characteristics and designs stabilize, those with adequate facilities and output should receive a comfortable income. Nonetheless, considerable engineering efforts will have to be made to maintain leadership, and lots of application notes written to upgrade the "old boys," who sadly need a full plate of saturated and unsaturated logic. Even your author is back in school learning the latest applications of commercial and consumer computers, because we live in today's world. Recommended are manufacturer's seminars, continuing college education classes, and field-engineering instruction. Sometimes, you'll need all the help available.

Time now to get on with the latest technology for the 1990s and bring you up-to-date as rapidly as possible. Therefore, Fluke/Philips becomes the next project, and they are, indeed, unique.

Fluke/Philips

Fluke/Philips ads introduce the new PM3580 family of logic analyzers (Fig. 6-4) which "give you state and timing" readouts "together on each channel with a single probe." You don't have to switch probes for either measurement nor change analyzer states for the two required modes. The PM3580 has 8 levels of timing, full state and timing information from the individual IC pin, and a price that's virtually half of its immediate competition. Channel acquisition ranges from 32 to 96 at 50-MHz state and up to 200-MHz timing. Glitch capture is specified as 2 ns plus 2K memory/channel for 8-, 16-, and 32-bit processors. Capacitive loading, even when both measurements are underway,

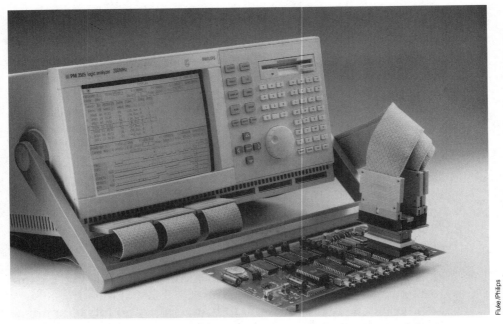

Fig. 6-4   *Philips top-of-the-line 200-MHz Model 3585 logic analyzer.*

is contained at 7 to 8 pF, while eight trigger levels allow 3 timing words as well as edge, time out, and glitch selections—a total of 8 state words.

State speed at 50 MHz should be rapid enough for many or most reduced-instruction set computers, in addition to transitional timing to extend memories. Per-channel memory is specified at 1 kbit for 100 MHz in the PM3580 and 2 kbits for 200-MHz PM3585, along with alpha/numeric keypads.

Fluke/Philips characterizes some logic analyzers as the ''instrument of last resort,'' because double probing interrupts technical thought patterns during the data time study, resulting in obvious time lost checking the instruction manual. Their pods, they say, continuously sample each probed line as state confirms the software function and timing reveals the 1s and 0s of the hardware.

As learning progresses, there are pop-up screens for various choices, MS-DOS for disk drives, and alpha/numeric keys. Three sets of keys set up functions, cursors, and controls; state and timing can be executed with a single keystroke.

In operation, push Run for timing, Select for state, and Insert for both. Then scroll through data to find whatever software or hardware difficulties exist for displaying or recording. Time and state triggers can be set to within 5 ns and you can zoom-in directly on the problem. All this relieves the puzzle of finding state data to pinpoint the trigger point, because it's already there. In these analyzers, program errors can be timed exactly and the fault serviced or eliminated—especially with the addition of a built-in floppy disk drive and Go/NoGo testing.

You might also easily acquire timing and data for all information and buses. When your state mode shows some difficulty, timing data is right there to determine the cause. If you have any confusion, pop-up menus can immediately aid in choosing the right path to rescue you from any misadventure. A single-trigger sequencer that operates over eight levels, including a possible mixture of state and timing trigger events should make these tasks even easier—especially with commonly used trigger sequences, which only require entry of the trigger words and the button-command Run!

Probably the best way to illustrate our previous highlights and discussions is to show you a pair of timing and state readouts with simple information called out for basic familiarity. The first example is shown in Fig. 6-5 from Analyzer 1 in the Timing Data mode. This mode was selected because basic oscilloscope and digital waveshapes are exactly the same as illustrated. The only extraordinary difference is the number of channels, level indicators and very helpful cursor lines (marked T, R, X, and S).

Let's explain: The name Analyzer 1 means initial setup, followed by a new acquisition in the Timing mode, accompanied by a time/division of 2 ms and then actual times of cursors R, X, and S. The operating mode is scroll, and the time difference between R and S is 4 000 ns. Trigger level(s) at S appear as 1s and 0s, because logic conduction occurs following the trigger point. Therefore, don't confuse trigger levels with values at the center frequency (cursor X), which, incidentally, is the inverse of level S. On the extreme left, C1KI are the channel designators (a, B, c, etc).

Patterns in Fig. 6-6 appear doubly confusing at first, but they aren't really difficult after a simple explanation points out their meanings and possibilities. Just remember, you're dealing with multiple inputs here, and their signatures must be recognized and instructions interpreted. We won't attempt to be precise in everything, but the fundamentals will certainly appear as useful information on which you can expand with use.

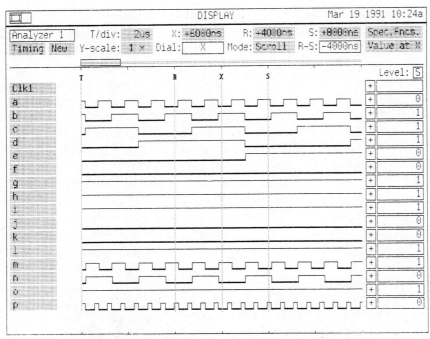

Fig. 6-5 *Analyzer 1 in the timing data mode.*

Label:	DSCTRL	FC2.0	ADDRESS	DATA	68000 Instructions		Time
Base:	+Hex	+Hex	+Hex	+Hex			Abs
-0001	4	6	000718	3415	MOVE.W	(A5),D2	- 175m
T 0000	4	6	000000	0000		RESET	+ 0n
0001	4	6	000002	b000			+1800n
0002	4	6	000004	0000			+3600n
0003	4	6	000006	0700			+5400n
0004	4	6	000700	4eb8	JSR	0430.W	+7205n
0005	4	6	000702	0430			+9725n
0007	0	5	000ffc	0000	mw		+14.4u
R 0008	0	5	000ffe	0704	mw		+16.6u
0006	4	6	000430	327c	MOVEA.W	#4007,A1	+12.3u
Y 0009	4	6	000432	4007			+18.4u
0010	4	6	000434	123c	MOVE.B	#98,D1	+20.2u
0011	4	6	000436	0098			+22.0u
0012	4	6	000438	1281	MOVE.B	D1,(A1)	+23.8u
S 0014	2	5	004006	98	mw		+27.7u
0013	4	6	00043a	4e75	RTS		+25.6u
0016	4	5	000ffc	0000	mr		+31.3u
0017	4	5	000ffe	0704	mr		+33.1u
0018	4	6	000704	347c	MOVEA.W	#4001,A2	+34.9u
0019	4	6	000706	4001			+36.7u

Fig. 6-6 *State readouts in hexadecimal might seem confusing at first.*

Here, Disassembly is On, Dial Y is identified at 0009, and the T, R, and S indicators are at 0000, 0008, and 0014, respectively (these are the selected cursor indicators). As you can see, trigger level T has been Reset, but is still on. The remaining connector contacts are shown generally in numerical order, except 0006, which follows 0008 instead of 0005 and 0013 (between 0014 and 0016); no 0015 is indicated. Go back to the column headings that follow DSCTRL, which represents disassembly (in most instances 4), then FC2-0 (the function code line). You will see a 6-number/letter Address column and a reset vector grouping programmed in at various points. T, the trigger, being Reset, is at 0 in this instance. In the next column, Data is registered as it moves along the bus at various logic positions. In the next-to-last column, the microprocessor 68000 instructions for movements and other commands are followed by Time in ms, μs, and ns.

The final combined state and timing readout appears in Fig. 6-7. Here there is no Disassembly, Y is shown as 0032 and it is the movable parameter identified by Dial, the R-S difference is

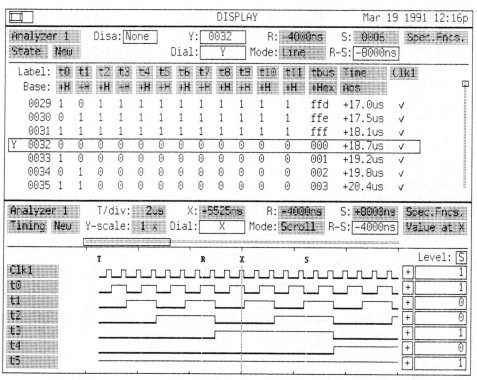

Fig. 6-7 *Dual state and timing modes make logic analysis simpler.*

8 000 ns, and addresses 0029 through 0035 are as shown. Timing for each lists between +17 and +20.4 µs and what data is there appears as 1s and 0s. Below in timing, the time/division is 2 µs, the X (center) cursor is +5523 ns, with R and S in their positions in the first of the timing figures. Displayed levels, however, have changed, and now total 6, and no cursor is identified "after" trigger, S.

Observe, however, that now only 7 channels appear in State and 6 in Timing. So, combining the two does restrict on-screen coverage somewhat, but the convenience is certainly worth some compensation when working in these dual modes. As indicated, the Y scale scrolls, so many parameters can be considered as the Dial moves the display here and there for additional channel inspection.

Remember, too, that in timing, you're working with the analyzer's internal clock oscillator that's asynchronous with the DUT. In State, it's the equipment's clock to which the analyzer must sync. For troubleshooting, known good clocks, subsystems, pc boards, or whatever can be recorded and retained for good/bad analysis.

This practice will undoubtedly become more prevalent as additional logic analyzers and more logic ICs and boards reach the market and are too expensive to discard. Therefore, you can expect that technicians as well as regular design engineers will be required to master the mysteries of this intricate, but highly necessary and useful equipment as very large scale integration (VLSI) proliferates.

Characteristics and definitions Philips/Fluke engineers do make some selected observations that might help to more readily understand limitations and advantages of all analyzers:

State speed This depends on the DUT microprocessor under test. Faster processors require more speed, but "clock-in frequency" might not be the operational or bus rate and often it can be lower. So, you'll have to know your specifications.

State memory Often a conditional specification, it identifies available memory, but remains somewhat dependent on trigger efficiency.

Setup and hold Setup time also affects recorded hold time. Together, they represent minimum signal stability for correct sampling.

Trigger levels   Any trigger level should acquire simultaneous timing and state information in a single setup and with one connection for half the work of separate systems.

Timing resolution   This is directly proportional to an analyzer's own clock speed. A 100-MHz clock, for instance, means 10-ns resolution, but it can reduce with multichannel use or special Glitch-capture function.

Triggering   Depends on available trigger levels and qualifiers, plus correlation between state and timing. Check specifications carefully.

Channel connections   Use high-impedance quick connects for best contacts in both state and timing probes.

Channel availability   Ordinary analyzers require channel assignments for either state or timing—often assigned in pod groups of 16 or, if demultiplexed, doubling channel grouping/signal. Obviously, this system is inefficient.

Data labels   Once defined, these are viewable in State or Timing, or both together in split-screen. Single trigger sequencers in 8 levels may be mixed and acquisition ceases when designated events take place. Multi-triggers are available from a list with trigger words entered and energized by the command, Run.

Self-test and de-skew   These operations occur on each power up and an extensive boundary scan test verifies the internal operation of each PM3580 instrument.

Dual sequencers   The PM3585 has one dual sequencer for each analyzer. In both state and timing, they can trigger two conditions in parallel and find the problem quickly with considerably less effort. State storage can select special words or conditions and exclude those not needed.

Logical separates   Another feature of the PM3585 is logical separates, where at any 16-channel section, you can split the equipment into two logically-separate analyzers; tracking, for example, the microprocessor with one multi-16-channel set and the remainder a 16- or 32-channel bus.

Such advantages have been pushed rather hard, but we feel that they are forerunners of even more simplified/sophisticated instruments now in design, which you'll be working with in the next few years. Whether they take this exact form, depends on ingenuity and requirements, but certainly they are setting a welcome trend of utility and ability, and do need more than ordinary circumspection. Perhaps a few demonstrations can convince, but that's your little red wagon. Our business is only to inform.

Scope or analyzer

Philips/Fluke has a worthwhile applications pamphlet on the ABC's of logic analysis, which should become even more useful as instrument usage is further understood and spreads. It makes some interesting comparisons between DSO and logic-analyzer (LA) functions, with additions.

Certainly, LAs would be used in program flows, glitches, interrupts, register loading, and even pulse lengths, but DSOs would be handy for line-noise examinations, some one-shot applications, analog functions, sine waves, some vector exhibits, and A/D/A examinations, etc. Thus your author would like to see more LA and DSO combinations, plus the significant advantage of single instrument timing and state analysis. This fond hope will be fulfilled in the very near future, probably by several of the larger manufacturers.

What the DSO can't do successfully is look at a number of digital lines in parallel, check design integrity, trigger at selected circuit points (even before the data), determine resolution by sampling rate, operate synchronously and asynchronously, and depend on its high digital accuracy for very close tolerance horizontal resolution. Conversely, a oscilloscope can read amplitudes, risetimes, waveform configurations, measure dc as well as ac voltages and, with the proper probes even measure current in terms of voltage-per-ampere or milliampere. Whereas the logic analyzer displays either a 1 or 0, but it can usually display in binary, hexadecimal, or another specific logic bases.

These instruments are quite different, obviously, but together they can deliver a great deal of information about the measurement world around them—so much so that a careful combination of the two at some fairly reasonable price would make them invaluable both on the bench and in the field.

State or timing

In the timing mode, a DSO also asynchronously samples one or two signals concurrently, while the analyzer can look at many. Analyzer readouts in this mode are produced in highly accurate timing diagrams in words, bytes, or data bit streams, bus activities, as well as timing delays or even hold states. Very high resolution timing diagrams will also display noisy transitions and line spikes of errant voltage (Fig. 6-8).

In the state mode (Fig. 6-9), the analyzer will sample signals whenever solid new data appears, usually using the system circuit clock of the equipment being investigated. Testee and tester are now synchronized and state lists of sampled data are immediately available. Each list supplies columns of data and the value of related information, the address or data bus, group status or special interrupt lines.

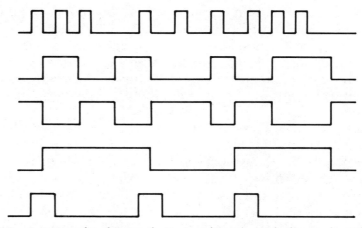

Fig. 6-8 *An example of a simple timing diagram with diverse logic on several channels.*

State	Bin	Adr	Data	Time
Trig	011	000360	BFCC	0 ns
00001	011	000362	522E	16.3 ns
00002	011	000366	66F8	40.2 ns
00003	011	000368	24C4	10.7 ns
00004	011	000380	F016	7.1 ns
00005	011	000371	4F10	36.3 ns

Fig. 6-9 *State analysis with trigger, address, data, and time (the latter is in nanoseconds).*

A state list, therefore (Fig. 6-9), is a record of valid, new data which delivers a considerably longer event track than that available from timing. Although state can never offer all the timing information, it can reveal the amount of clock pulse trigger detection, but it can't show logic instability. So, any wrong values will appear just as though they were good ones. Consequently, state and timing information should always be simultaneously available for accurate data examinations, whatever the equipment or source.

Design errors and hardware operating conditions remain in the prime use for logic analyzers, although complete equipment examinations are often too extensive for single tests. Therefore, piecemeal verifications are usually basic applications, although oscilloscope investigations are needed to pickup such problems as crosstalk, ringing, risetimes and falltimes, ac amplitudes, dc values, pulse symmetry, and the like.

LAs predominate in manufacturing partially through automation and selectively on pc boards, with good/bad comparison bit streams probably already filed on reference disks. Obviously, the pressing need for logic analyzers will increase substantially as digital logic (commercial and consumer) equipment becomes more complex and analyzer programming becomes somewhat simpler.

Philips/Fluke says that maximum recording lengths and high-resolution factors are essentially solved by transitional timing, which means that storage only occurs with changes in state. That is, zeros to ones or vice versa. Actually, such information stores in both data and time memories. If channel data changes between samples, older information stores in data memory, while time memory collects valid data. Meanwhile, storage efficiency increases as sampling increases, but loses as inputs become faster. However, because recording lengths don't depend on sampling frequency, the analyzer always samples at maximum resolution. Transitional sampling, then, offers extended recording, maximum resolution, and avoids the necessity of manually regulating an internal clock. So, efficiencies of 250% or more over conventional timing are said to not be unusual.

Triggering and clocking

Glitch-catching is usually the prime sport of logic engineers, who have about 5 ns to recognize their quarry and often need latches or pulse expanders to do the job. Philips/Fluke says that one way divides memory into a pair of sections, one of which stores glitch and the second stores the timing diagram. Unfortunately, the recording length is halved, and the glitch is not specifically identified. Another method involves adding an additional channel to each ''signal.'' However, this method halves the available channel numbers, so both attempts are somewhat unsatisfactory.

For triggering, built-in logic can often recognize some special circumstances and it can be programmed to execute at some particular high/low transition. Trigger delay can also be used and even pre-trigger and post-trigger conditions can be examined. There might also be more than a single trigger word and certainly the timing of a pulse length, whether shorter or longer than normal by the edge or sample period, and timing analyzers can have more than one trigger word.

State analyzers accept samples from the DUT's own clock and change cycles accordingly. If such samples are strictly those needed in monitoring, only relevant data is stored, which permits longer record lengths and increased capture possibilities. Therefore, an analyzer might need to track several clocks, its own and that of the DUT, and you might have to decide which you want to be involved.

Consequently, the number of channels in state analysis will probably determine those in any logic analyzer; these often depend on the subject equipment's logic analyzer. Here, full sync between the equipment and the analyzer is required before strobing so that data properly fills the analyzer's memory. Usually, any sample-and-hold timing should be under 40 to 50% of the clock cycle. A 100-ns period, for instance, will require sample and holds of 20 ns for accurate tracking. The number of channels depend on the DUT's processing unit, plus extras for control, interrupts and addresses. For exact numbers, consult your Philips/Fluke dealer or other reliable source. Otherwise, even a slight mistake will cost equivalent accuracy and probably cause suspect and incomplete readouts.

In state analysis, the trigger can operate on either edge or level triggering. On edge, triggering only occurs when incoming data changes to that of the trigger. In level triggering, the analyzer compares data with selected trigger words and operates upon match. Level triggering guarantees a positive reaction, and edge triggering responds immediately upon any data change. Philips/Fluke has no special preference.

Data acquisition and analysis

Storing the proper data effectively uses the state analyzer's memory. Called *relevant data*, a clock qualifier masks clocks during irrelevant data inputs; one or more clock qualifiers are needed for each separate clock. Their use, however, can add to setup time, because clocking information passes through extra gates and causes delays.

Selective data logic, however, stores only certain samples so that the analyzer has maximum use of its memory, which benefits the operator considerably.

Timing diagrams are much like those that might be read on a multichannel digital scope, but they are impractical as a result of not only the number of channels, but also the special triggering, bus activity, channel indicators, highly specialized commands, restricted amplitudes, and often, extended frequencies. In short, LAs and DSOs appear to be similar, but they are two considerably different instruments.

For clear identification, the channel number and signal name must be displayed and placed at convenient locations, usually above or below one another. When the signal flow propogates along a bus, such might overlay the individual signals that identify various transitions or other phenomena, then the bus information is erased at will.

For special measurements, cursors are available for pulse characteristics or even the length of some pulse trains. These can also be moved at will by keystrokes from the operator.

State lists are the companions of timing diagrams, but they provide very different information. Here, you want to know exactly how a circuit operates.

Therefore, when external clocking begins, all analyzer pods are sampled during triggering, labelling channels 1 through 9, followed by Status, Address, Data, and even an indicated time (in ns), which tells the event times. All this then shows line/channel activity and permits reading of data, where it's stored, and other information that is pertinent to logic operations. Also, some microprocessor examinations include instruction fetches and eventual reconstruction of the assembly code for information readback. Lists that include reverse assembly instructions are usually called disassembly lists.

There's much more to these simple timing and state coding schemes than we've portrayed here. Sorry to say, logic analyzers aren't exactly simple in the state format—especially when each instrument has a separate protocol of its own. Therefore, choose carefully when making your selection, because you might be working with it for some time to come.

The most important characteristics are timing resolution, sufficient channels, clock and trigger speeds, memory size, state and timing words, clock numbers and qualifiers, special glitch detection to approximately 3 ns (if possible), probe impedances and close state and timing relations.

Disassembly

Probably the only category we haven't dealt with at some length is *disassembly*, which becomes highly useful for microprocessor-controlled circuits and systems. Information is reasonably easy to observe in ordinary testing, but it's vitally necessary to comprehend and follow programs that are delivered by the microprocessor.

Disassembler instructions come in special packages that consist of a floppy disk with appropriate software and a microprocessor adapter. Such adapters are designed for a minimum number of pods (probe holders) so that many other signals can be accessed and displayed also.

In the PM 3585, adapters are designed for maximum processor timing data capture, including both state and timing information. Simple to load, the disassembler is merely inserted into the analyzer. Press Select on the Option field, and a disassembler list appears on the CRT, which displays all available routines. Then, highlight a particular disassembler and press Select. Pods are designated label and clock assignments are given in Format, prior instructions are queried and, if accepted, you're ready to decode (unless you'd like to change Parameters in the State list). Select will also offer a popup menu for any different disassembler parameters that you desire.

Now you should have a reasonable "feel" for microprocessor decoding and be ready to work both hardware and software for reasonable and very specific results. Don't, however, expect miracles if you don't have an adequate logic analyzer. They are miraculous, but like computers, they are not yet thinking machines. You must tell them what to do and stay within their individual limitations. Good luck!

❖ 7
Time-domain reflectometry

TIME-DOMAIN REFLECTOMETRY AND TIME-DOMAIN TRANSMISSION are pulse and measurement techniques that your author first became aware of in 1974, (or possibly a couple of years earlier). At that time, Russ Harding of Hewlett-Packard wrote a piece for *Electronic Design* entitled "Use Pulses Instead of CW Signals to Analyze Discontinuities in High Frequency Circuits."

Since then, additional articles have been written on the subject, but few have explored in depth the whys and wherefores of TDR/TDT, along with the developing technology that has made such instrumentation so very useful in both metal-cable and fiberoptic analysis. Therefore, although such techniques are not wholly those of an ordinary oscilloscope, their resulting displays and specific measurements are shown on a cathode ray tube. In this book, that's the basic qualifier. With this qualifier in mind, we'll see what a few contemporary sources can contribute. The results will be passed along to you in, we hope, reasonable substance and continuity.

As Mr. Harding stated, TDR/TDT is a pulse-reflection method that is often compared to a closed-loop radar system. Accordingly, a fast-rising step (one level to another) of voltage propagates down a length of cable or electrical system at some specific impedance and at a certain velocity. Any abnormality that is encountered causes some portion of the step to be reflected and appear as a discontinuity in the transmitted waveshapes. Where this blip or suckout occurs is easily timed and recognized so that both the distance from the point of origin and the type of problem can be dealt with almost immediately, which provides access to the system if the cabling is open, rather than buried or closed.

Basics and more

Distance to any such discontinuity is:

$$D = Tt_o/2\sqrt{e_r}$$

where T is the speed of light, t_o is the time between step and discontinuity, and e_r is the relative constant of the dielectric.

For air, the equation simplifies to:

$$D = Tt_o/2 \text{ or } D = T(t_2 - t_1)/2$$

All this equation signifies is that an exceedingly fast pulse risetime is necessary for precision measurements. Indeed, this statement is true because low ns only measure low-value meters, but minimum generator risetimes are needed for small millimeter resolutions, which would reveal difficulties with connectors, terminations and other hardware that operates at high frequencies. A table of step values clearly indicates some of the forthcoming probabilities and the relationship between line-terminated impedances and resistive loads—$Z_L = 2 R_o$, as an example (Fig. 7-1).

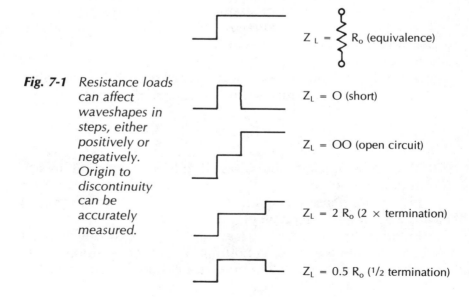

Fig. 7-1 *Resistance loads can affect waveshapes in steps, either positively or negatively. Origin to discontinuity can be accurately measured.*

$Z_L = R_o$ (equivalence)

$Z_L = O$ (short)

$Z_L = OO$ (open circuit)

$Z_L = 2 R_o$ (2 × termination)

$Z_L = 0.5 R_o$ (1/2 termination)

Observe waveform reactions according to termination, which could also become line or systems problems (somewhat more subtly). Notice that for resistive loads that exceed Z_L, discontinuities are positive. For those that are less than Z_L, discontinuities are negative ($1/2\ R_o$, for instance). At the moment, we'll not overwork these simple examples until more of the TDR theory has been explained. After that, we'll see what can be done with real instruments that make the theoretical a reality. Just to suggest what occurs when resistance replaces impedance, transmission line theory states that:

$$P\ (\text{rho})\ =\ \frac{Z_L\ -\ Z_o}{Z_L\ +\ Z_o}$$

All of which means that Z_L is the load impedance, Z_o is the discontinuity's impedance, and ϱ (rho) is the voltage reflection. In very short lines, Z_o usually remains resistive.

For line aberrations that are strictly resistive, any reflection looks like a portion of the incident (transmitted) voltage. However, when dealing with RLC reactive impedances, circumstances are quite different—especially at varied frequencies where capacitance can be predominantly reactive in slow transmissions, and inductances are the prime factor in speed. This is why the "skin-effect" of cable and cladding in fiberoptics is most important in evaluating transmission lines; RF and data seek the path of least resistance.

We can also generally assume that capacitive problems induce a positive-going charge; inductors do the opposite (a negative exponential). Now, we know that no real ideal transmission lines exist because losses result from resistance, leakage, etc. Therefore, to account for both shunt and series losses, the input impedance of any transmission line can be evaluated by:

$$\text{Zin}\ =\ \frac{R\ +\ j\omega L}{G\ +\ j\omega C}$$

if R, G (conductance), C, and L are all in terms of per unit length (Equivalent Fiberoptics equations are explained later). Operator "j" represents phase change in 90° increments and ω (omega) means $2\pi\ f$ (6.28 × frequency).

Entering further into the real world, the impedance of an ideal cable should appear as nothing more than a straight line. Problems within or along this line are anything but straight, and some are alarmingly crooked.

For instance, observe the inverted waveshapes in Fig. 7-2 with 10 × expansion. The upper one shows a small capacitance spike somewhere along the cable, and the lower one simulates a drop in impedance, which could result from poor braid cover and a secondary voltage intrusion somewhere along the line. Remember, also, that conversion from 50 to 75 Ω involves a 5.72-dB loss if your TDR isn't double-jointed (switched). A number of considerations exist in doing this type of troubleshooting, where any sort of impedance changes are involved. In the event of multiple discontinuities, you might want to identify and eliminate them one at a time, beginning at the transmitting-signal origin and working forward. Where only small impedance changes occur along the line, the effects are not large and can often be overlooked, especially if their locations are difficult to service.

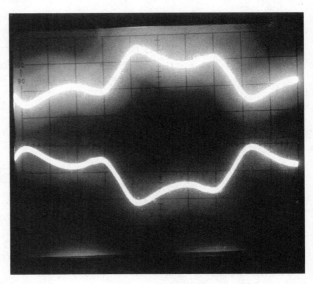

Fig. 7-2 *Inverted and normal waveshapes (10× expanded) illustrate discontinuities on nonlinear transmission lines.*

The waveshape in Fig. 7-3 can be caused by too-close coupling between generator and cable under test; it might well indicate poor quality or mismatched inductance cable. If this waveshape occurs, couple signals from the TDR or system under test with several feet of air line or high-quality cable. If the ringing doesn't go away, try another piece of similar cable and see if the same pulse delivers equivalent results. If not, you have the answers. If so, check the terminations, as well as the competing lengths of cable. Then, lower the input frequency to see if it changes. After that, connect your TDR directly to the oscilloscope. Your TDR shouldn't lie!

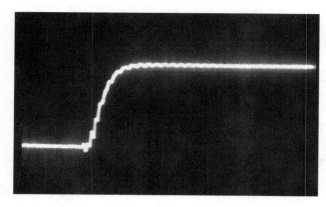

Fig. 7-3   *A classic example of high-frequency ringing.*

Using basic hookups with TDRs, cable impedance and loads are easily measured, delay line attenuations and retards, ringing, frequency characteristic, and other ordinary and special measurements are all possible when good instrumentation and wideband oscilloscopes are available. If capacitors are in series, discontinuities will increase upward from normal, but increase downward if in parallel. The reverse, of course, applies to inductors. Although pulses are extremely sharp for measurements less than 100 ft., pulse widths, which usually appear as a half sine wave for greater distances, are lower frequencies and can measure as much as 1 000 μs, (1 ms) in width. Actually, all this is done with different equipment which is discussed as we continue. Other equipment, services optics (fiberoptics). We'll have much to talk about there, too. Hopefully, little or nothing will be left to chance if and when you tackle your first TDR—or even wrestle with a new one, where applications are probably more refined and somewhat different.

Complex loads

As you have already seen, resistive loads only result in step functions that are obvious. Unfortunately, real-life terminations are often more complex than not. They require special considerations, which are not exceptionally difficult, but do need close attention and a known constant or two. Doing such exercises the hard way involves Laplace transforms and I doubt if either you or I want to experiment with many of these. Easier methods do exist, however and we proceed with those next; remember that inductors initially buck a current change, then later tail off to zero. Conversely, a capacitor begins its charge immediately and exponentially, the charge rises in a positive direction. The inductor, however, resists current changes as an infinite impedance before beginning its equal and opposite curve to that of the capacitor. Response to series and/or parallel RC/RL combinations performs similarly. So, what we're really doing is evaluating the reflected voltage when $t = O$ and $t = $ infinity.

If a transmission line was totally lossless (we've never heard of one that is, except possibly with superconductors), the impedance Z in ohms would equal the square root of its inductance divided by capacitance:

$$Z_O = \sqrt{L/C}$$

where L is in henrys and C in farads.

However, matched transmission lines do represent nonreactive loads to the signal origin (source). Thus, impedance doesn't change without the introduction of problems, and unit-length LC components do not contain pure resistance. Any divergence in wire size or spacing does affect inductance and capacitance, respectively, which, in turn, does affect impedance. So, you should not monkey with either balanced or unbalanced "lumped-constant" transmission lines.

Whenever transmission lines are charged over specific lengths, they develop an altered characteristic impedance that is resistive, rather than reactive. Both reactive and resistive components cause energy propagated down the line to reflect some portion of the incident energy. If physical changes in either balanced or unbalanced lines are necessary, they should occur in increments over short distances to prevent sudden changes in Z_o. A coaxial cable, for instance, that is abruptly switched from 50 to 75 Ω, develops an impedance difference of 5.72 Ω, which is not only a loss, but the lost energy would be reflected back toward the signal source as a capacitance mismatch. Smaller or larger cables and quality coupling will reduce such reflections considerably, if not entirely.

We have already said that relatively short-length coaxial cables have a complex impedance that includes dc, skin-effect transmission, and shunt conductance that is:

$$Z_o = R + j \omega L/G = j \omega C$$

Conductance G in mhos (representing shunt conductance) is normally negligible in coax and it is not often used in the equation.

Conductor skin losses can be minimized with clean and smooth-surfaced cable. Further, transmission-line RLC elements are evenly distributed over short lengths and appear as a simple capacitor at the lower frequencies if the line is unterminated and its electrical length is shorter than its wavelength.

Keeping this information in mind, you can now determine the characteristic impedance of a line if you know its dielectric constant (k), the outer conductor inner diameter (D), and the inner conductor's outer conductor (d). The dielectric constant helps determine the capacitance per unit length and so it is very important. Therefore:

$$Z_o = 138/k \times \log_{10} D/d \qquad \text{(138, of course, is a constant)}$$

In addition, dielectric k also affects the signal velocity of propagation through the cable because electromagnetic energy would travel at the velocity of light (or 300 000 meters/second), if unhindered. Through solid polyethylene dielectric, this velocity slows to about 0.659 of light speed, which is considerably slower. A good RF cable today has a propagation velocity of 78%, and an excellent cable will pass energy at 82%. Also, the per-foot inductance of this cable relates directly to the two conductor diameter ratios.

Now, any transmission line that is terminated in resistive Z_o is said to represent an infinitively long line. Otherwise, you will induce reflections, which is what complex loads, terminations, and couplings are all about.

Test signals

Fast pulses are used in TDRs to test some transmission lines, but not over the same bandpass that is calculated from the 10-to-90% risetimes. So, no TDR tests transmission lines quantatively as a separate system with sine waves. Often, the two methods are complementary, but a very fast pulse also possesses extra energy at extremely high frequencies. On the other hand, an equal sine wave has an amplitude that is unequal to that of the pulse. These are interesting concepts and we'll expand on them later in the chapter.

Coax often induces a step-function that changes the originating signal. Long cables induce greater distortion. However, it's known that they sometimes attenuate some high frequencies less than lower ones. Therefore, long cable step responses are often measured between 0 and 50% risetime points, rather than between 10 to 90%, under normal circumstances.

Short cable sections, as previously indicated, are virtually lossless and are measured accordingly. So fast/slow risetimes must be applied judiciously for appropriate results. This is especially true of cable sweeps where the front and back ends of long cable can be seen, but not the center. Consequently, TDR examinations are really the only way to go if you want to see responses all the way. Return losses on good-quality swept cable, by the way, vary between 15 and 30 dB, depending on computer/video/satellite-application designs.

A useful little equation, called the *voltage reflection coefficient* rho (ϱ), is derived from the total display amplitude, less the incident pulse amplitude. So, $E_{reflected}$ divided by $E_{incident}$ (the driven voltage) becomes:

$$\varrho = E_{reflected}/E_{incident}$$

and this can be expanded to include $R_L - Z_o / R_L + Z_o$, which also equals ϱ when the transmission line terminates in resistance equal to its characteristic impedance.

Resistive discontinuities produce different reflections: one is *step* and the other is *changing*. Series transmission-line resistances result in positive reflections whereas parallel resistances induce negative reflections, which emphasizes once again the extremely important aspect of exact terminations. As we continue, you will find that the signal generator's impedance must also match that of the line, but only the reflected signals are ever actually evaluated. If tests are for length only, however, don't worry about special deflection factors; only propagation velocity is really of prime interest.

Large reactances

Small reactances are often overlooked, but the larger ones need to be included here so that you can become familiar with their induced shapes and general characteristics. Thanks to Tektronix, we have examples of capacitors, inductors, simple equations, and reactive responses. Transmission lines might not exhibit precisely the same effects, but you can at least recognize some of the important characteristics by knowing how RLC circuits react as loads and time constants. The term ''large reactance'' doesn't mean one of great magnitude, but one of more than normal duration. Therefore, if an L or C lasts for 63% of 1 time constant, then, in TDR terms, it is large. Usually, however, a TDR incident pulse provokes only a short-time reaction and appears differently in testing. So, these reactions are considered separately and will be illustrated by "live" photographs, rather than the usual time-constant approach.

The information in Fig. 7-4 is typical for capacitance line terminations. The first is a series output that is terminated in a shunt capacitor, whose equation becomes:

$$C = \frac{1T}{2Z_0}$$

and possesses the step and exponential curve shown.

Next, Z_0 is in parallel with the capacitor so that the equation becomes:

$$C = 1T/\frac{Z_0}{2}$$

Circuit	Equivalent circuit	Formula	Display
Series with terminated line		$C = \frac{1\,T}{2\,Z_0}$	
Parallel with terminated line		$C = \frac{1T}{\dfrac{Z_0}{2}}$	
Across line end		$C = \frac{1\,T}{Z_0}$	

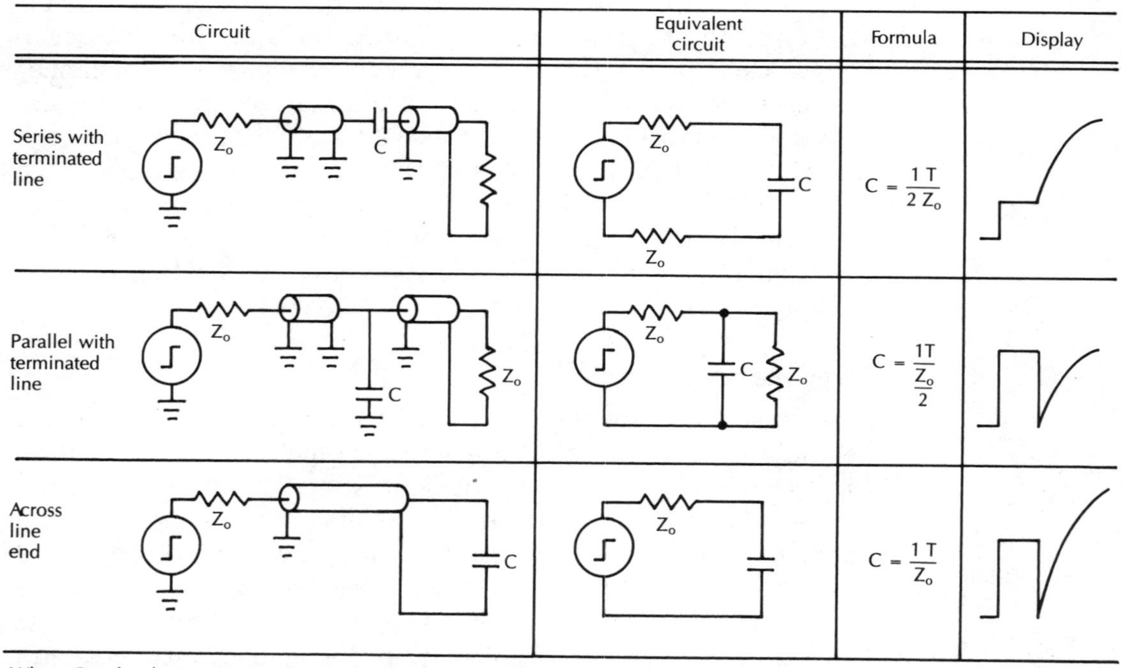

Where C = farads; T = time constant; Z_0 = line characteristic impedance

Fig. 7-4 *Large capacitative reactance line terminations.* Tektronix

which permits first a pulse, then a modified exponential rise.

Finally, a single Z_o in series with the terminating capacitor also allows the pulse, but its following charge curve is both sharper and higher:

$$C = 1T/Z_o$$

As said before, inductive reactions are somewhat equal and certainly are opposite from capacitors for the quick- and slow-charge reasons given. Figure 7-5 really represents current through an inductor, opposition, and then leakoff of the charge following the usual reaction. Consequently, series Z_o and L induces the following value of L:

$$L = 2Z_o \times 1T$$

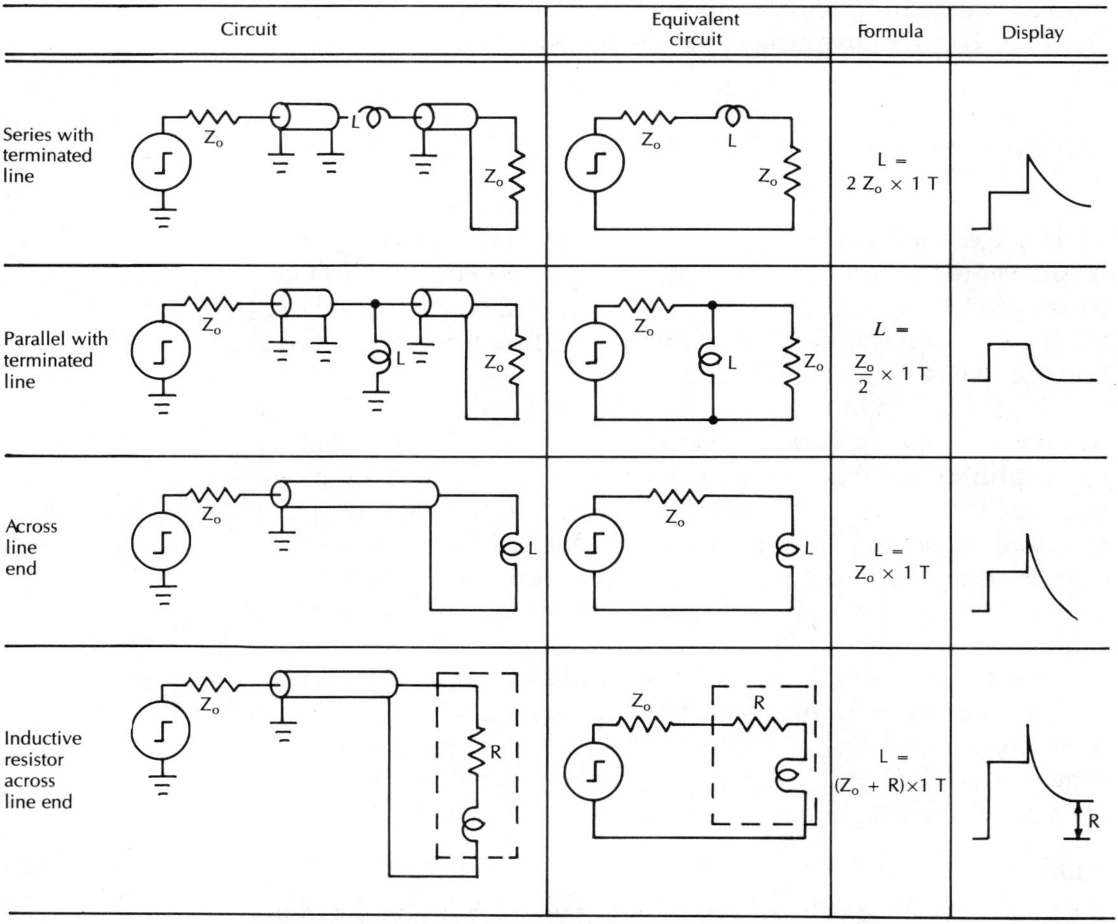

	Circuit	Equivalent circuit	Formula	Display
Series with terminated line			$L = 2 Z_o \times 1\,T$	
Parallel with terminated line			$L = \dfrac{Z_o}{2} \times 1\,T$	
Across line end			$L = Z_o \times 1\,T$	
Inductive resistor across line end			$L = (Z_o + R) \times 1\,T$	

Fig. 7-5 *Large inductive reactance line terminations.* Tektronix

A generator in parallel L and Z_o equates to:

$$L = Z_o/2 \times 1T$$

With a simple inductance in series with Z_o, the equation is:

$$L = Z_o \times 1T$$

An inductive resistance across the line becomes:

$$L = (Z_o + R) \times 1T$$

but offers only a small portion of the decay which levels out at R above the full decay as in a simple L termination.

Small reactances

If you were dealing with a small-series inductor, with L (in henries) and Z_o (in ohms), the value for L would be:

$$L = 2.5 \, \alpha \, Z_o t_r$$

The equation for a shunt capacitor's value is:

$$C = 2.5 \, \alpha \, t_r / Z_o$$

The Greek letter α relates to the reflection coefficient, or ϱ (rho). If you wanted to compute the charge for a 50% capacitor/inductor time curve, it would be 0.693 of any one constant. After this, we'll await some actual demonstrations before we continue with both charges and decay.

With the foregoing very basic TDR approaches, you should, by now, be ready to understand the real meat of the chapter: both in amplitude and time resolutions—the actual measurements you really conduct with your TDR. *Limitations* include noise over longer sections of transmission line and noise plus risetime over short sections, which now probably involve pc boards and/or other small-system circuits in measurements that affect amplitude.

Risetimes are principally the fundamental problem with time resolution. However, the fastest TDR with the least noise does not necessarily become the most suitable, as we've stated previously. Different instruments serve different purposes, according to what they're measuring.

Padders

If you are unable to calibrate your 50-Ω generator so that its impedance matches that of the line it's pulsing, an L-pad can easily do the trick—just keep it contained in some nonradiating enclosure to prevent leakage (Fig. 7-6). Illustrated is a typical 50- to 75-Ω conversion, which is always handy, but be prepared for a signal reduction of 5.72 Ω as a result of shunt and series dividers. In addition, because no capacitors are in the network, this unit will

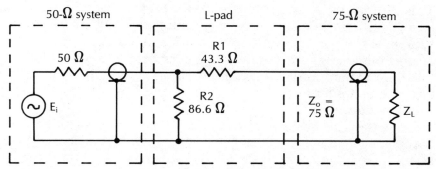

Fig. 7-6 *Typical 50 – 75-ohm resistive matching pad. Be aware of the 5.72- dB power reduction. The voltage varies little.*

pass dc. You must be especially careful when connecting any such circuit between equipment that is sensitive to dc, particularly most spectrum analyzers. If you do not have the exact values calculated for this or any other compensator, either additional series or shunt resistors can be added and measured with a good digital voltmeter that has at least three-place accuracy.

$$for\ Z_o\ greater\ than\ 50\ \Omega$$
$$R_1 = [Z_o (Z_o - 50)]^{1/2} \quad R_2 = (50 \times Z_o/R_1)^{1/2}$$
$$for\ Z_o\ less\ than\ 50\ \Omega$$
$$R_1 = [50(50 - Z_o)]^{1/2} \quad R_2 = (50 \times Z_o/R_1)^{1/2}$$

Circuit-board connections/connectors

The following information is primarily furnished by Tektronix, because this company's a very major supplier of TDRs and is always generous with its applications and technical information. Although not verbatim, we have tried to work closely with both expression and intent; you should find the content more than useful.

For circuit-board connections, fixtures or quality probes are required for accuracy and repeatability. Connectors cannot change between measurements, nor can permanent fixtures (such as "bed of nails") be moved, because even small position or connection shifts can produce considerable changes in impedance. Nonetheless, any applied TDR connection should result in repeatable measurements; this is a good way to check your immediate results.

When using probes, try to mount it (or them) in a permanent, rigid fixture, with a probe securely held in one position. This setup is especially appropriate for large production runs or special circumstances where maximum accuracy is required. Where less-accurate numbers will suffice, the special probe with correct impedance and bandwidth can be firmly "impressed" on the circuit board under constant pressure (without body contact) near the pc under test. This procedure could require some interesting dexterity.

Tek says that alligator clips or similar connectors and nonimpedance-controlled pins are not recommended because they often cause a considerable impedance mismatch and are not suitable for repeat connections. Several illustrations graphically demonstrate how poor-quality probes and clip-lead connectors distort waveforms that would otherwise appear considerably more linear with proper terminations.

The following three examples (all from Tektronix because of the tiny distances and only connector/terminator involvement) illustrate how sensitive a short-range TDR must be for accurate results. All three are taken from actual printouts that follow displays and storage via the cathode ray tube.

In each instance, the strip line is merely 4 inches long, but seen in Fig. 7-7 as a controlled impedance measurement with a 50-Ω probe, a 50-Ω cable, and a 50-Ω termination. The connector causes the usual rise at the beginning of the measurement, but everything settles down quickly as a fast pulse traverses the line. At 0.2 ft/div. and 40.9 mp/div. the beginning and ending of the

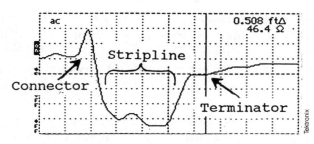

Fig. 7-7 *Evaluating a 4-inch stripline with 50-ohm probe and a 50-ohm termination.*

display are on the same plane for a total delta change of 0.508 ft. and a readout impedance of 46.4 Ω at propagation velocity of 54%. So, the actual, measured impedance is within 90% of its projected value.

In Fig. 7-8, a 2-pin connector essentially causes havoc at the stripline-connector junction and delivers such poor vertical resolution that the readout registers only 45.2 Ω at 59.4 mp/div. Here, the connector is obviously the prime problem and has forced the trace entirely off the screen. You don't want this type of display.

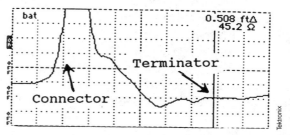

Fig. 7-8 *The pin connector causes impedance mismatch and poor vertical resolution.*

In Fig. 7-9, the reaction is not terrible with the clip-lead connector, but impedance difficulties remain for some period following the connector. Resolution becomes miserable, especially the vertical.

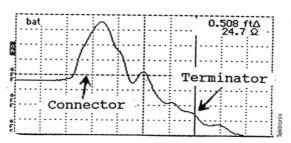

Fig. 7-9 *Clip lead illustrates unacceptable distortion in both vertical and horizontal measurements.*

In choosing either BNC or SMA connectors, BNC offers a 4-GHz bandwidth versus 9 GHz for SMA. So, choose wisely and solder or crimp well. As you can obviously see, probes and connectors are absolutely important in such short, stripline, general pc, or other constricted-length examinations. It doesn't take much to upset the cake tray.

Try to always choose the proper connector and make the best possible securing crimp. Connectors or terminals have three prime areas: electrical mating for current carrying and contact resistance (including strength and plating), wire size and conductor's crimp (to specifications, if possible), and current return shielding. An application tool should crimp the termination firmly—not enough to mash the shielding through its dielectric nor so loosely that the termination disengages. Production equipment usually secures connectors according to specifications. For simple hand-held crimpers, just remember the instructions and examine each crimp with a short test, an open test, and a fairly rigorous pull test. We always use a digital voltmeter before connecting relatively short cables. If there's multicabling, a dc voltage at one end and a VOM at the other should identify the ''hot'' line. Just make sure that you're operating on an open line and not one that is already connected to some $1,000 computer processor.

LONG LINES TDRs

This series features Tek's model 1503C, which came down from Beaverton, Oregon for use and evaluation. From this study, both of us will learn things and the writeup should have a foundation of authority. First, let's look at a few specifications and features.

The 1503C (Fig. 7-10) is described as a long-range, metallic cable tester that is useful on coax, twisted pairs, and parallel cables. Impedance changes along any cable are reflected as assorted ''hills and valleys.'' The instrument can find shorts, opens, shielding problems, accumulated water, kinks, and many more ''hidden'' faults by generating a half-sine wave, and detecting and processing any return voltage from the incident (transmitted) signal. This waveshape can be temporarily stored and recalled or printed fully on a dot-matrix printer that is contained as an option in one dedicated port. Operating power derives from ac or an 8-hour dc battery, and the 1503C with battery, cover, and recorder weighs 20 lb.

The instrument is calibrated in volts over time, just like an oscilloscope, but with selectable pulse widths of 1, 10, 100, and 1 000 ns. In distance measurements, its range is equal to or greater than 50 000 feet, with a resolution of 0.04 feet at an accuracy of 1.6 inches (± 1% of distance measured). Velocity-of-propagation settings (V_p) are from 0.30 to 0.99 at resolution of 0.01 and an accuracy of 1%. Pulse output impedances can be selected

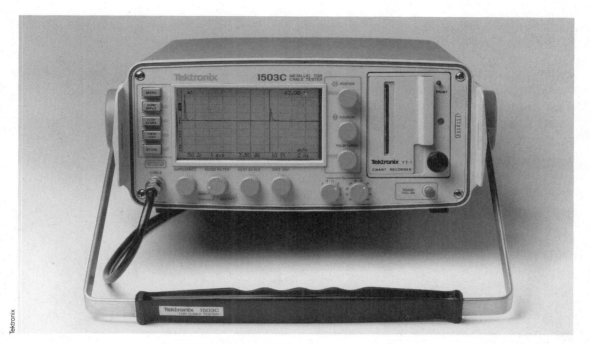

Fig. 7-10   *Tek's excellent metallic cable tester with accuracy of 1%.*

at 50, 75, 93, and 125 Ω within 1%. Vertical scales can also be read from 0-to-63.75 dB gain(s), and incident pulses are settable within ± 3%, with terminated pulse amplitudes of – 2.5 V and unterminated at – 5 V for 10 100, and 1 000 ns. Random displayed noise at less than 1 div. peak can be shown at 63-dB gain.

This information should sufficiently give you an idea of the instrument's capabilities. Next, we move into setup and applications, which are surprisingly simple, considering 1503C's internal complexity and accuracy. Once you've learned the initial adjustments and have worked through preliminary measurements, you'll probably feel like a veteran.

TDR Setup and applications

Called *cable radar* by some (with good reason), these TDR's are menu-driven instruments, which suppress noise by the addition (or subtraction) of averaging noise with noise filters (caused by long cables that act as antennas and absorb surrounding electrical discharge). Systems consist of a microprocessor, a gate array, an address decoder, and memory interrupt logic.

An 8-bit microprocessor clocked at 5 MHz addresses inputs, decoding, and memories (Fig. 7-11) up to 65 536 memory locations. A video processor contains a vertical-position D/A converter, a summing amplifier, a video amplifier, and an A/D converter. Instructions are read from the EPROM (electrically programmable read-only memory) by the microprocessor for most instrument functions. The RAM (random-access memory) below permits the processor to store, retrieve, and control video and display information as programmed. Address decoding circuits read/write data or triggering-controlled functions. Interrupts reach the processor via these decoders, requiring data reads for source interrupts, status, followed by service routines. A 20-MHz clock generator provides divided-down system timing for 5-, 2.5-, 1.25-MHz and 625-kHz frequencies for both digital and analog time bases.

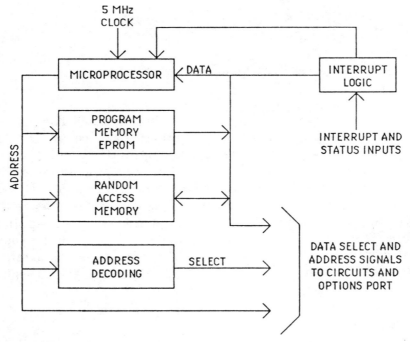

Fig. 7-11 *The heart of the 1503 is an 8-bit microprocessor with memory and decoding.* Tektronix

Setups can either be gleaned from the instrument's CRT, printout menu (Fig. 7-12), or from the service manual/instruction book. You will see explanations of each instrument control;

→ Return to Normal Operation
  Help with Instrument Controls
  Cable Information Menu
  Setup Menu
  Diagnostics Menu
  View Stored Waveform Settings
  Option Port Menu

Move ⬍ Position to select, then Push MENU button

Fig. 7-12 *An actual equipment printout which offers menu instructions for those who need guidance.*

cable aids with impedance and velocity of propagation values; a setup menu for maximum hold, pulse on/off, single sweep on/off, vertical scale, distance/division, and background light on/off; a diagnostic menu for LCD dot pattern test, response time, driver test, contrast adjust, LCD chart, head alignment, diagnostics for sampling noise and impedance, offsets and gain, and time-base settings; stored settings for waveform display; and an option port for signal verification, timing, and debugging.

That should be sufficient for starters and casual acquaintance. Now let's begin working on combined setups and applications that should see us through the metal TDRs and on to the optical ODTRs that follow. Be aware, however, that the vertical raster is set up in either decibels ($1/10$ of a Bel) or millirhos (the ratio of reflected voltage divided by the incident voltage as $1/1000$ of a rho). Single division pulses measure 1 000 millirhos, two divisions are 500 millirhos, and 4 divisions would signify 250 millirhos. Horizontally, the 1503C calibrates from 2 to 1,000 ns. Take your pick.

Just to become acquainted, a 3-foot cable without termination was used as Tek's suggests for the initial setup and to become acquainted with the instrument. As the printout in Fig. 7-13 shows, this is an open-circuit adjustment so that the reflected wave is almost the amplitude of the incident wave and the cursor is set at the midpoint of the graticule. This distance is a little less than 30 percent of the reflected's rising edge, with a readout of exactly 3 feet, and the distance/div. shows 1 foot. Pulse width here is 2 ns and the impedance is exactly 50 Ω. Do you have an idea what would happen if the cable was terminated in a perfect 50 Ω? No energy would be reflected. In this setup, you can move the cursor wherever you wish, but both incident

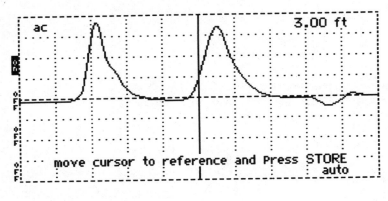

Cursor .... 3.00 ft	**Fig. 7-13** *Setting up*
Distance/Div ..... 1 ft/div	*with a 3-foot*
Vertical Scale.... 6.25 dB	*precision*
VP 0.66	*cable. The*
Noise Filter ...... set Δ	*cursor is set*
Power .... ac	*at 3 feet.*
Pulse Width ...... 2 ns	
Impedance .... 50 Ω	

and reflected voltages remain the same without movement. Therefore, know your cable lengths when making measurements or measure them electrically before continuing.

If your cable length is unknown, set Dist/div. to 5 000 feet and work backwards by decreasing the distance adjustment until the reflected pulse becomes visible. Short pulse widths depend on cable lengths, and you might have to both increase the vertical scale setting as well as increase the horizontal pulse width.

Here, we really didn't know either the length of this cable or its actual impedance, although the latter is usually printed somewhere on every cable's jacket. Between 50 and 75 Ω, the trailing edge on the incident pulse became either shorter or longer (75 was longer), but an obvious difference was at 93 Ω and, therefore, 75 Ω was selected as the proper setting. We played with both the pulse width and the distance knobs and found that 2 ns and 2 ft/div. were adequate.

Observe that the reflective pulse is actually negative (Fig. 7-14), but don't let that throw you. There's a good reason for this: We deliberately pressed the View Diff button and the return pulse appeared as a negative, but somewhat lower in amplitude. Is this correct? Hardly, because there is no stored pulse and, therefore, no difference. What you are seeing is only the reflected pulse

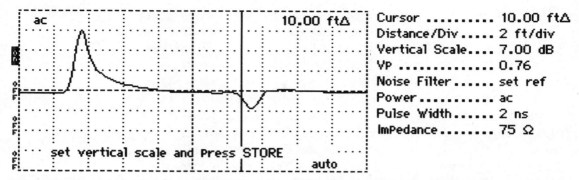

Fig. 7-14 *A cable with unknown length and impedance. The reflected voltage appears negative.*

with a negative dip and no incident pulse at all. Setting the Distance switch to 5 ft/div. proves the point. Confused? This problem can easily happen if you push too many buttons. Stick with the basics first!

The printout in Fig. 7-15 is actually what you should have seen had the normal readouts been referenced. Both pulses are now positive, although number 2 has a little less amplitude as a result of internal cable resistance. In the two waveforms, however, the vertical scales differ, but all else is normal, unless you regard 0.08 feet as an error. At any rate, playing with unfamiliar buttons doesn't always aid the cause.

Let's do something more with Fig. 7-15 while we're in dB. The 5.25-dB readout is the measure of the reflected voltage and not that of the incident. Secondly, if you raise the amplitude of the reflected voltage to that of the incident (or reference), then you find the return loss of the cable, which as 7.75 dB in this instance.

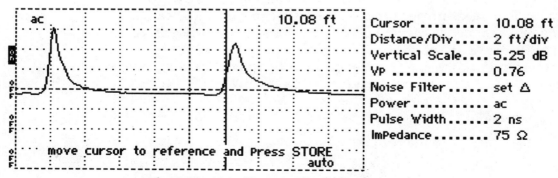

Fig. 7-15 *How Fig. 7-14 should have looked.*

Figure 7-16 can now become somewhat more interesting because some of the basics are fairly well established. We're going to insert a small pf capacitor (actually a spectrum analyzer dc block) between the 10-foot and the 3-foot cables, couple the two cables with a 50-to-75 Ω impedance matcher and view the result. As you can see, the cursor reads 13.84 feet following that curious negative- and positive-going bump, which is 10 + feet from the incident pulse. If you didn't already know, this distance means that some sort of capacitance/inductance reaction, which occurs over a space of 1.6 feet between the two newly connected cables would require attention under any circumstances. Now, you have an idea of how sensitive this equipment is and how it locates and measures faults with consummate ease. Don't forget that a large rise in the return waveshape is always an open; a small rise or fall is usually an impedance fault.

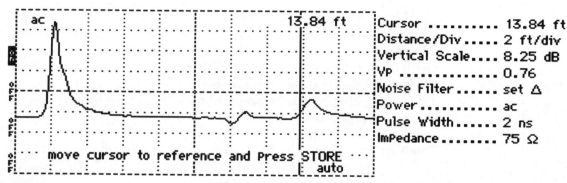

Fig. 7-16 *Insertion of a capacitor and impedance matches between two cables does make a difference.*

There's more to the vertical scale than just dB. Rho (ϱ) is next. This measurement is also calibrated in millirhos, which denotes the 1503's vertical sensitivity. Conversion between dB and 1/1000 of a Rho appears easily with the simple press of a front-panel button. Waveshapes in the previous figure, for instance, are now 273 mϱ. This figure is really the power delivery of the cable or system. It is the ratio of reflected voltage divided by applied voltage, and it is actually the system return loss in dB.

$$dB \text{ (return loss)} = E_{return}/E_{incident}$$

As you already know, an open circuit returns almost all incident voltage as a positive pulse. However, a short generally returns energy of lesser impedance than the cable as a negative pulse. One measurement is from 0 to + 1 and the other is 0 to − 1.

Therefore, each scale division from 0 equals a maximum 1, two divisions equal 500 millirhos, three divisions equal 333 millirhos, and four divisions equal 250 millirhos. So, above or below 1 division, the millirho values decrease from unity or 1.

Millirho

To demonstrate using the millirho, we'll switch to 50-Ω cable in a considerably longer length (measured at 23.67 feet). This cable is RG8/U with polyfoam insulation, which, according to the printout in Fig. 7-17, has a V_p of 78%. Velocity of propagation, of course, determines much of your measurements, and (in common with millirho) selects the instrument's readout sensitivity of the reflected image. Any other setting, especially a false one, will immediately produce inaccurate information. If you have none of the cable characteristics other than 50 or 75 Ω, go back and recalibrate the TDR, take a good guess at the dielectric, then try and match an open line reflection with that of the incident in shape and approximate amplitude. After that, you can begin whatever measurements are of interest. Later, with usage, you'll become more scientific, but right now we're primarily eyeballing.

```
Velocity of Propasation (VP) depends mostly
on the insulation (dielectric) in the cable.
Jelly Filled ......................... VP = .64
Polyethylene (PIC, PE or SPE) ...... VP = .66
TFE or PTFE (Teflon) .............. VP = .70
Pulp insulation .................... VP = .72
Foam or Cellular PE (FPE) .......... VP = .78
Semi-Solid PE (SSPE) .............. VP = .84
Air (helical spacers) .............. VP = .98
       - Push MENU button to Exit -
```

Fig. 7-17 *The 1503C prints out this useful listing upon menu command.*

With all of the above accomplished, the cable length in Fig. 7-18 is exactly 24 feet and the dielectric-contingent velocity is a proven 78%. Now, add another section of RG8 cable, but hook up a dc block in between, just to offer an idea of how higher frequency cable might behave in TDR examinations. This cable does precisely what you might expect (Fig. 7-19).

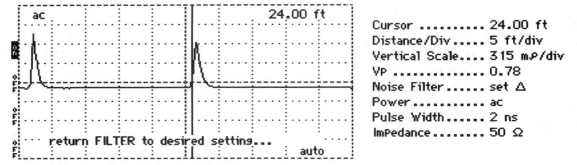

Fig. 7-18 *The open-ended RG8 measured roughly 23.67 feet (actually, it's 24 feet). Close enough?*

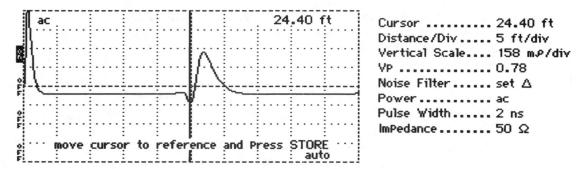

Fig. 7-19 *A stray impedance is usually very visible and easily measured.*

The amplitude of the return pulse is more than a couple of millirho down from incident as a result of the 50-Ω termination, and an interesting negative-going bump is at the 24.4-foot mark. Isn't that our dc block? It most certainly is, and we also know that additional cable has been added, but the following reflection doesn't really show us exactly how much. At any rate, the negative pulse of interest is of a lesser impedance than the cable, although not a short circuit, and it is reflected back (as shown) in opposite and greatly reduced polarity.

Once assimilated, you should have had a pretty good workout with TDRs on cable measurements. Your author is firmly fascinated with them. We can't, however, continue this chapter forever on metal-measuring TDRs. So, the next section will deal with ODTRs, but in considerably less detail because reflectance technology does have kindred similarities. The exception is that instead of driving an incident signal down a wire, the incident wave is now a light source.

Optical TDRs work much like metal-step TDRs, except that they operate on light pulses and a detector, rather than electrical stimulus. Actually, the three basic types of ODTRs are: Fresnel, Backscatter, and Fault Finder—all of which we'll work through individually or collectively as the material unfolds.

Optical reflectometers

Fiberoptic transmissions can be either analog or digital, depending on intended objectives, short- or long-haul considerations, and applications specifics. All signals must be modulated and demodulated; the results steadily improve as a result of fewer glass impurities and vastly updated electronics. Signal leakage in fibers is nearly nonexistent; therefore, it neither radiates nor absorbs. So it is an ideal transmission means for virtually all data—especially because many fewer amplifiers and vastly less maintenance are needed; much wider bandwidths than metal cable can be used; exceedingly rapid transmissions can occur; and it has a high resistance to noise, temperature, and humidity. It's somewhat less practical for transmitting low-capacity carriers to a large network (such as CATV), and it is currently more expensive. With the arrival of wireless cable and production/ price ratios, there's little fiberoptics (FOs) can't do in the future versus metal cable. If AT&T—our favorite telephone company (we didn't necessarily say long distance, where several other systems are competitive)—can talk the Federal Communications Commission into stringing FO into homes, you might (and probably will) see some radical changes in consumer services in the U.S. (at least across North America, perhaps the world). Consequently, this portion of the TDR/ODTR explanation is of considerable importance to both signal propagators and recipients, not to mention to the thousands of engineers and technicians who are involved with laying and maintaining these lines.

Although this chapter is not devoted exclusively to ODTR operational aspects, we do believe that you should understand the whys and wherefores of optical cable before we launch into 1990s techniques and testing. It's our contention that unless you know the guiding principles of system operations, all the instruments in the world won't permit intelligent troubleshooting and signal analysis—both of which are essential for satisfactory results. Therefore, we'll undertake a modestly rapid explanation of fiberoptics cabling and what makes it a probable next step toward the eventual development of low- or no-impedance super-conductors that operate near room temperature.

FO specifics

With the complete ground isolation between receiver and trans-mitter and immunity to electromagnetic interference, one might think that no other ODTR problems are of significance; this rea-soning is wholly false.

In order to operate, fiberoptics systems must have a drive cir-cuit, an emitter, connector(s), the cable itself, a detector, and an amplifier. The FO has a center core of material, with a high index refraction that is capable of carrying the optical information and a "cladding" layer with a lower index of refraction that sur-rounds this core for fiber isolation and external/internal emis-sions containment. Materials require extreme purification and excellent finish on the core surface to prevent signal absorption and light scattering. These are the two prime attenuation factors and are usually measured in decibels/kilometer at some special nanometer light frequency. Attenuations can range from 1 000 dB/km to much less than 50 dB/km, depending on plastic or high-quality glass cores and/or cladding. If more than one cable is jacketed, the combination is said to be *bundled*, and a cable with a large diameter that is relative to its operating wavelength is called *multimode*. In short, the principle of fiberoptic cable operation is one of accurate internal reflection. Its aperture (Fig. 7-20) calculates as:

$$\text{Sin } \theta = (N_1^2 - N_2^2)^{1/2}$$

N_1 is the core, N_2 is the cladding, and θ is the maximum angle that light can be propagated between the core and the cladding. The positive fit of couplings, of course, is critical for successful cable operation. If the system features multiplexing and demulti-plexing for extra information transmissions, then insertion loss, crosstalk, and channel bandwidths are sensitive parameters. Also, although *encription* can increase bandwidths, it might induce more than normal attenuation.

Single-mode fiber has lower joint and coupling losses as well as less attenuation, minimum Rayleigh (light) scattering and greater resistance to gaseous hydrogen and nuclear radiation at

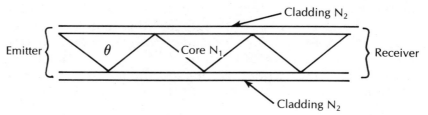

Fig. 7-20   *Principles of fiber cable core and cladding.*

lower cost. With multimode cable, the bandwidth seems to saturate at about 2 GHz/km. Single-fiber transmissions of analog optics above 2 GHz now seems promising. Development of direct laser-diode modulation and traveling-wave indirect modulation are aiding the process considerably, and the recent development of fiberoptic absorption losses to considerably less than 1 dB/km is a real breakthrough. Microwave fiberoptic applications are now turning more and more to injection lasers, rather than LEDs because the latter is less efficient and not as linear as lasers, which are coupled with very low noise and fusion-spliced optical fiber sections, where there would be ordinary connectors with their usual discontinuities.

Common fiberoptic detectors are often photodetectors, which produce a current that is proportional to the optical radiation intensity. Their efficiency is characterized in terms of ampere/Watt(s) all linked to speed, noise, and physical compatibility. So, the detector's area must be approximately the size of the optical fiber's core. Thermal and shot noise can be problems. Most common detectors are PIN diodes and avalanche photodiodes.

Critical speeds have specific wavelengths that you should know because high speeds are assigned 1 300 to 1 500 nanometers, but shorter wavelengths operate between 700 and 900 nanometers. As wavelengths increase, so does cost. Low-data-rate systems are limited by power loss or fiber attenuation, but faster systems contend with pulse spreading in the fiber. Typical wavelengths are 665 nanometers for plastic fiber; 820, 850, and 870 nanometers for glass fiber in shorter wavelengths, and 1 300 and 1 500 nanometers for long wavelengths. These are tested by power meters that measure optical-fiber power in milliwatts, microwatts, and dBs (from 1 μW to 1 mW), and optical time-domain reflectometers. These reflectometers sense fiber backscatter characteristics to locate faults such as nonmatching impedance, poor couplings, return losses, problem fibers, and they can be matched to wavelength and fiber core diameters for maximum measurement accuracies. Measurements, however, must be considerably more than fairly accurate. At 1 300 nanometers, for instance, a good single-mode fiber has a power loss of 0.8 dB/kilometer.

Connections

Source material for this portion was gathered in late 1989, but it should still offer worthwhile information for present and past systems that continue to use such cables and their dressed and undressed endings. True, certain improvements, such as auto-coupling devices, are now on the market, but these are probably a rarity in all but 1991 or later systems. This information is included in the hope that it can be of considerable value for older and even current optical cabling.

Optical defects and losses are minimized by careful connector matings, where splices or couplings are. To do this, their ends must be physically and optically flat, free from any surface pits and scratches, and no air gaps can be in between.

When connectors are not necessary, cleaving, mechanical splicing, and fusion are all approved methods of joining any two ends. Exact fiber alignment allows the least insertion loss, and rotary splicing actually tunes the two cables for best insertion loss.

Both manual and automatic means can polish fibers. In manual polishing, hand-held fixtures and several abrasive polishing operations are involved (in addition to experience and skill on the part of the polisher). Inexperience will often injure or actually rupture part of a cable. In autopolishing, the fixture doesn't contact the abrasive area during the process. However, the process still involves a series of graduated texture abrasives until the fiber surface finally becomes smooth and flat and free from damage. The final step is the application of epoxy, its type, and the method of application. You might find that porous alumina ceramic connector ferrules are more difficult and time consuming because of fracturing the alumina when removing old epoxy. Our source recommends a 1-μm diamond compound with a liquid extender that is applied to a napless polishing cloth.

The standard throughout the industry for bare and connecting optical fibers includes: a 3-μm silicon-carbide abrasive bonded to a poylster disc/sheet, a 0.3-μm aluminum-oxide abrasive bonded to a polyster disc, and either a water or liquid abrasive extender. Fiber sensors often require polishing to special angles when used by medical personnel for pressure measurements and temperatures.

Installations

Most of these involve little more than investigation and common sense. However, fiberoptics are extremely sensitive to dust. Don't leave raw, cut ends exposed to air without a cover a tape or a cap, watch temperature specifications, and don't permit glass to penetrate the body, because X-rays can't see it. Twisted or broken cable always means damage. Otherwise, maintain the same taboos that you learned when working with metal cable. Nearby machines, vehicles, ladders, ditch diggers, etc. are constant hazards; don't stretch the cable until it stresses or breaks under strain and the buffer coating ruptures.

Once plans have been formalized as to cable placement, minimize splices, connectors, pull loads, review all handling procedures, and try to install connectors after the cable is in place. Coatings can be removed either mechanically or with an approved chemical solvent.

As for tools, you'll need a pulling eye or grips, cable lubricant, and a pull-force monitor. Splicing requires a special enclosure, a knife, a jacket stripper, a cleaving tool, a splice unit, cleaning pads, and other appropriate materials. For attaching connectors, use cable strippers, a fiber-cleaving tool, cleaning and polishing paraphernalia, a power meter, and an optional source.

ODTR applications

Backscatter ODTRs look at Fresnel and reflections detected by the Fresnel method. This is similar to the metallic TDR scopes, except for the addition of an optical-electrical converter. It (or they) are capable of high-resolution measurements to fiber ends, as well as mechanical connections, but it requires a Fresnel-fiber reflection to do so.

Such reflections can be seen by Backscatter instruments, as well as by small amounts of light that are reflected from glass melts (fusion splices) at various connections (Fig. 7-21).

Fault Finders are actually Backscatter ODTRs that evaluate reflections to local connections and total cable lengths in dis-

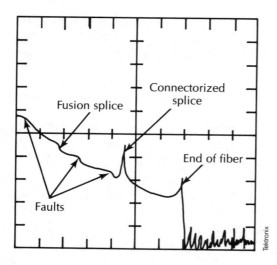

Fig. 7-21 *An example of fusion and connector splices, as seen by an optical time-domain reflectometer.*

tances, rather than waveforms. In short, these are the main types of optical time-domain reflectometers, with which we're primarily concerned. Let's detail some information on exactly what they do.

ODTRs are usually equipped with a pulsed light source, such as a laser diode, a photodiode, a cathode ray tube, comprehensive electronics, and a selection of optical parts. Cables are connected to the *launch port* (optical output) of the ODTR, which emits a series of light pulses that are usually in the 800- to 900-nanometer wavelengths. If the optical fiber is both perfect and infinite, no light reflections result from such incident injections (Fig. 7-22).

Fibers are not often perfect conductors, and they are normally terminated. Splices, connectors, couplers, etc. are often prevalent throughout their lengths. All of these anomolies can be traced and are usually identified virtually precisely, because each incident pulse generates a series of optically "visual" returns that can be detected via the ODTR by measurement and often by positive signature. Very little guesswork remains if you're working with a first-class ODTR.

Tektronix supplies the following equation to calculate fault distances, which includes both constants and variables:

$$L = c/n \, (\Delta t/2)$$

The distance between incident and echo reflections (L) involves "n" at 1.47 for the fused silica step index, or (1.45 for fused silica with graded index): Δt, the time lag; and 3×10^8 meters/second

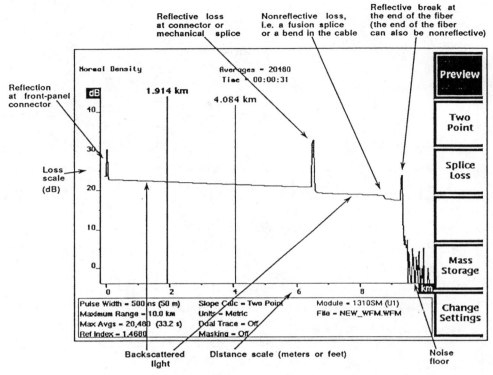

Fig. 7-22 *Tek's TFP2 newest Fiber Master sees it all, even in preview.* Tektronix

c, as the velocity of light in free space. Δ/2 represents the "round-trip" reflection.

A Fresnel Reflection occurs as light passes from one glass core to another, which has a different index of refraction. Forward traveling power, reflected by an air gap, is identified by the equation:

$$\%\text{Pref.} = N_{\text{Fiber}} - N_{\text{air}}/N_{\text{fiber}} + N_{\text{air}}$$

N, of course, is the refraction index, as shown above.

Rayleigh Backscattering, a second reflection mechanism, is the prime loss factor in better high-grade fibers. Any equation for this phenomenon is both difficult and complex, it exhibits according to Tektronix, a decaying exponent, with an amplitude that is proportional to input power times pulse width.

All of this makes it possible for ODTRs to accurately show cable characteristic and faults in terms of vertical Y-scale amplitudes and reflected time delays on its horizontal X axis (Fig. 7-23). However, overly strong incident signals into fiber cable can

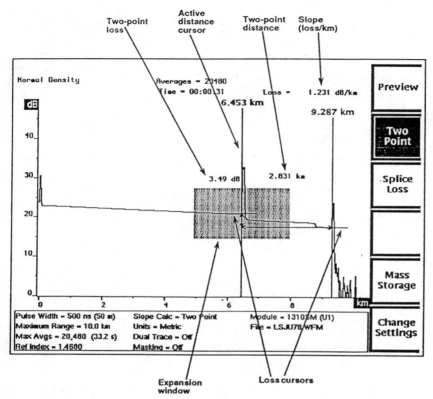

Fig. 7-23 *Two-point mode calculations reveal maximum information in fiberoptic cable examinations.* Tektronix

result in ODTR detector overloads, which negate any possibility of measuring problems in close proximity. Ideally, fibers being tested should deliver high Rayleigh loss and a substantial index difference between the fiber core and the surrounding cladding layer; neither characteristic is good for high-quality F/Os in distance measurements. Therefore, fiberoptic measuring instruments must sense such characteristics and compensate accordingly. Not only this, but their input sensitivities should certainly have isolation levels that approach or exceed 50 dB. Good signal levels and sampling rates are of the utmost importance for better-than-average results. A 30-dB isolation isn't considered acceptable any more—especially with the recognition of unusual problems.

Bent fiberoptic cable, for instance, could be tapped! In security and business communications, this situation is serious and it requires careful signal analysis to positively discover. In some cases, a detector that is placed next to the cable can actually receive signals through its jacket; the smaller the bend, the better the detection. Previously, average fiber power only could detect gross taps. Now with very stable instruments, changes in average power can now tell when fiber(s) are tapped. When incident light reaches a bend in the fiber, a critical angle occurs that will not permit total reflections that are needed for maximum signal passage. At this point, some of it might or can escape through cladding and jacket and be picked up by the pirate detector. Any drop in average power, therefore, alerts the operator to the possibility of evident problems. This means a high degree of vertical resolution sufficient to display a trace akin to a fusion splice (Fig. 7-24).

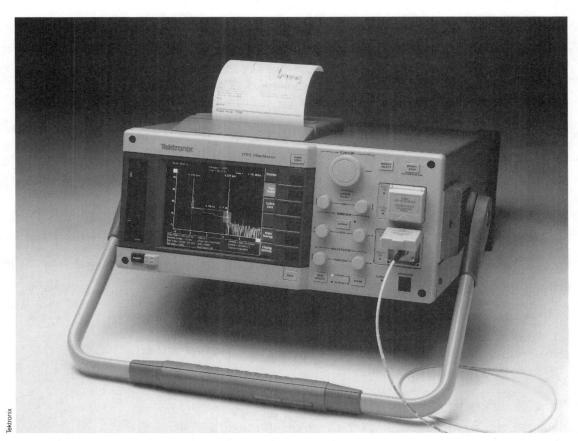

Fig. 7-24 *The TFP2 Fiber Master ODTR tester with permanent record printout.*

Definitions

The foregoing, naturally, are generalities, which were mostly included for both instruction and applications, where appropriate. They do, however, suggest very different terminologies from other types of measurements, of which you should be aware:

Dead zone really represents detector saturation, during which accurate backscatter measurements are impossible because full recovery from this effect, including pulse width, amplifier magnitudes, and detector "tailoff" are ongoing and incomplete.

Readout resolution is displayed as dB change. High-loss resolution allows greater detail on screen.

Data point spacing is spacing between trace points, with points in exact positions corresponding to those on the fiber. Closer data points mean better resolution. Keep these points in the low centimeters.

Marker resolution varies from meters to kilometers (more or less), depending on your instrument. Low meters to high kilometers are always desirable under distant conditions. Pc-board examinations are different; here, distance is not the prime specification, but close-in accuracy is. Tektronix' OF150, for instance, has a dead zone from 6 to 8 meters; a readout resolution of 0.1 at 2 dB/div. or 0.5 dB at 10 dB/div.; data point spacing of 2.5 cm; and marker resolution of 1 meter to 1 kilometer. Designed for multimode fiber, the OF150 can measure cable up to 4000 meters in length. Other ODTRs have difference characteristics, of course, so choose both accurately and wisely for your own particular applications.

If your requirements are distant, rather than short range at very modest cost, you might consider a TFS2020 FiberScout™ from Tektronix that offers resolutions of 1 meter or 1 foot, loss measurement resolution of 0.1 dB, fault threshold selections of 1 to 4 dB, a repeatability of ± 0.3 dB, and a fault detection range of 11 dB. Readouts are numerical, rather than strictly CRT, but for wavelengths of 1300 nm at long range and 835 nm at short range, this equipment is hard to beat.

Valuable tips

Because you have a reliable display or printout with TDRs and ODTRs, why not make a profile of each new or "known-good" installation, and save it for reference and posterity? If a recorder isn't connected, snap a Polaroid for on-screen charts and figures, and file it away as a standard for yourself or for your successor.

Surprisingly, such measurements can apply to both cable and antennas. Cable testing has already been explained, but *transmit towers* are another story altogether, although similar tactics prevail. Because a Tektronix 1502C can identify problems within 0.004 feet you can even see where cables enter and depart from connectors. The 1503C will resolve faults separated by only 11 inches, so it is normally used for any of the usual tower and transmission-line problems. Even if the cable is wet *and* has a connector problem, you'll be able to spot each difficulty separately for prompt repair. Reminder: A wattmeter can't "look at" discontinuities, and certainly not water.

If you are plagued with *intermittents*, the 1500C has a Max Hold feature that continuously monitors the cable without erasing the previous trace. Shake and pull the cable and connectors all you want, rattle the antenna, move the cursor to the occasional fault, and send your technician skyward for the fix. If your difficulty occurs only in a high wind, however, be considerate and have repairs consummated during a lull. Tower sway is sometimes intimidating, to say the least, and confidence can easily be shaken.

Occasionally, you might have a wire or *wirebond connections* between circuits or pc boards that induce discontinuities, especially inductive. Sometimes those discontinuities can be attenuated by a pair of capacitors on either end of the normal inductance. If you're working with 50 Ω, let's say, and you can measure the small wire reactance, the following equation can help:

$$Z_o = \sqrt{L/C}, \text{ with } C \text{ being } C_1 + C_2$$

Don't forget to square Z_o to obtain values for L and C.

Telephone faults are additional problems for which the 1503C is well suited. However, a different approach is usually necessary because of the special characteristics of phone lines. Therefore, instead of beginning at the phone terminal, working between load coils often produces maximum results. Although we offer a rather compact explanation of what occurs in this and other methods, you are happily referred to Tektronix' 22W6347, which describes time-domain reflectometry application for telephony. Numerous waveshapes and detailed explanations are available within, including a very useful graph on coax and communications-grade twisted pairs.

Tek says that ordinary fault locators are usually of two types: *terminal* and *tracer*. Both have problems with operational sensitivity, awkwardness, and limited abilities. Tek recommends a TDR that is guided by plate maps in the hands of a trained technician to quickly locate and identify all fault types, on any cable, twisted pair or coax.

This prime reason is why you are being referred to Tek's publication, rather than a somewhat long-winded paraphrase of the same material. If you purchase a 1503C, request this invaluable publication also.

Basically, half-sine wave pulses of various widths can range to 50 000 feet on quality coax, and to 35 000 feet on #19 AWG cables. Phone "line" signatures result from tip, ring, and sheath problems when there are faults.

Open tips and *rings* produce the familiar positive reflectance(s) that are common to all cables; tip-ring shorts are negative "valleys," an open sheath induces another positive blip, but a shorted sheath does just the opposite. Line inductive load coils ring positive, but water initiates a low tail-off positive/negative integration effect for the duration of the "wicking" or water ingress.

The foregoing is just an introduction to what the 22W6347 contains. Active telephone testers should become thoroughly familiar with its contents when using the 1503C. You'll be glad you did!

❖ 8

The laboratory of tomorrow

"MAYBE I'M RIGHT AND MAYBE I'M WRONG," AS THE OLD '40s tune goes. Nevertheless, the omens are propitious and the Ides of March must be near. This time, let's turn about and praise Caesar, rather than bury him. Brutus should get lost!

After watching stand-alone instruments grow in ever-increasing complexity, system demands have increased beyond the call of simple analysis and repair, and vast amounts of treasure have been spent to maintain adequate design and system checkouts. It's plausibly evident that the day of software control of virtually everything necessary for most if not all of these needs is actually (almost) at hand. Just a little more software and a 4-foot diagonal flat screen pasted on the wall is all that's left for our national and international inventors to produce. Meanwhile, of course, we'll probably navigate through the era of the super-brilliant flat-tension mask tube (now in production by Zenith) and some smaller and duller liquid-crystal or molecular offerings (by various and sundry). In the end, the eye-savers and dedicated troubleshooters will arise from the debris to supply us all with the electronic sights and numbers that will move us into the 21st century. Displays of all descriptions would then be digitally stored in super-archive disks for future reference. The programmers could then turn their attention to the next series of marvels that require any and all programs for execution and reactive operations.

As you are all aware, unattended robotic "Men of Mars" constitute a considerable population of our heavy-industry factories, and more are on the way everywhere that their utility is practical. They work longer, harder, and with less downtimes than humans and a 5-day work week is nothing more than a warmup; except, perhaps, a pause for lever and software messaging and lubrication.

So it must be with engineering aids and de-glitchers: a keyboard, lots of storage disks, a flat screen, and thou (thou is the equipment setup for electrical laundering).

No hangar flying

To illustrate that this isn't the old blue-sky mirage, many instruments now feature IBM, Epson, HP-IB, and other formats basically for printer outputs and other instrument-to-instrument talk. The latest interfaces that are available in more sophisticated instruments are the IEEE 488 (known also as GPIB); EIA's RS-232; and VXI. Plug-in data-acquisition boards can also be used to turn the computer into an instrument. These standard interfaces already permit instruments to "talk" to one another over matched impedances and equivalent interconnects. This "talk" allows rapid and accurate movement of information from one port to another with minimum loading and maximum speed. The relative attributes and preferences for one format over another aren't really our concern, but we will briefly describe their general outlines in an ensuing section.

As a result of their very existence and dedicated operation, the hour draws much closer to our concept of the panel and the keyboard. At the moment, some equipment has one or two of these interfaces (usually the 488 and the RS-232), but no more. However, one manufacturer has supported all four in his software and hardware compositions, not only for the U.S. but also for 13 countries abroad (as of June 1991). This company will host over 30 two-day seminars across the U.S. during 1991, and more thereafter. Titles are "Data Acquisition Systems for Measurement and Control" and "Instrument Systems for ATE and Advanced Test and Measurement." Your author will attend both sessions and will duly report their contents. The company, National Instruments of Austin, Texas 78730-5039, has mainframe and associated equipment displayed in Fig. 8-1.

Already, says National, there is a Graphic User Interface (GUI) that combines a graphics display and text-based interface,

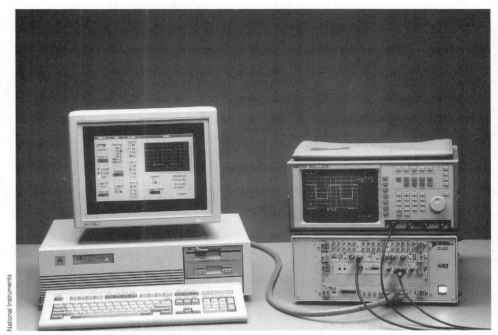

National Instruments

Fig. 8-1   *A good example of National Instrument's mainframe hookup with Hewlett-Packard and Zenith equipments.*

which not only displays information, but also prompts an operator for inputs. With GUI, the system developer can design a software panel that becomes a full-function instrument. Consisting of graphical panels and pull-down menus, they are often mouse-driven so that a mouse-cursor can manipulate data in familiar and straightforward directions. Such GUIs, generated with Lab-VIEW 2 and LabWindows 2.0, can produce pushbuttons, switches, sliders, menus, and graphs, plus shapes, sizes, colors, etc. Data applications, such as data logging, process control, biomedical and physiological sciences, will be useful for managers, engineers, and scientists. At the very least, we shouldn't have microfische to worry about forever, or until we physically and/or mentally go blind.

As for storage, magnetic and floppy disks, RAMs, ROMs, bipolars, and CMOS are already with us. Magneto-optics show promise in both medical and satellite imaging in the form of magnetic large-density dipoles, which can even be rewritten when the occasion demands. Obviously, a wealth of technology is available, and once instrumentation, remote programming, and direct, high-speed testing is acquired and understood, all of us will proceed to higher levels of system examinations with considerably improved information and troubleshooting abilities. It is already suggested that test and other instrumentation procedures can be recorded and distributed to national and international facilities for comparison with valid and nonvalid DUT readouts. Isn't that an improvement over crumpled schematics and flickering shadows on the wall? You might even have years of manufacturer's information on a single floppy disk. As you can obviously see, design and service data are already intertwined for many engineering people of today and the future.

GPIB and RS-232

Good promises become positive realities, not ''goomby,'' as retired Congressman Morris Udahl used to say when describing hazardous barnyards. So the business of outstanding communication buses comes first, then National Instruments® features and operations follow (Fig. 8-2).

The GPIB, or IEEE-488, is described initially because National Instruments (NI) advertises it as the leading supplier of GPIB interface boards. Its latest boards are IEEE-488.2 compatible, letting you take advantage of this standard as well as Stan-

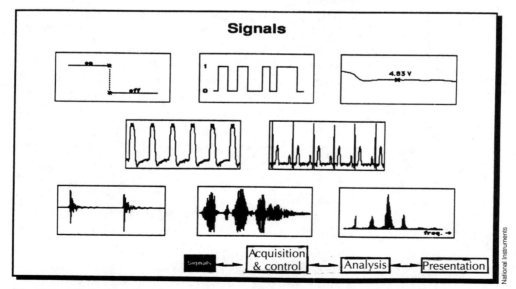

Fig. 8-2 *Electronic signals can be virtually anything. The trick is to find and display them.*

dard Commands for Programmable Instruments (SCPI). GPIB represents General Purpose Interface Bus.

ANSI/IEEE-488 (1975), now known as IEEE-488.1, simplified programmable instrument interconnects by "clearly defining mechanical, electrical, and protocol specifications," according to NI so that instruments from various manufacturers could be equally connected by a single-source cable. IEEE-488.1 did not include standards for status reporting, data formats, common commands, error handling, or controller requirements. IEEE-488.2 (1987) does all this and more. It decidedly reduces instrument development periods by precisely defining how controllers and instruments "talk" to one another, and it even permits considerable interaction with 488.1 devices.

The IEEE-488.2 requires the controller to include an exact and complete set of IEEE-488.1 interfaces, capabilities, bus control sequences and protocols, and documentation standards. The controller must also possess the ability to sense the transition of the SRQ signal line plus the level of this signal, and detect and report time outs. *Protocols* are high-level routines for test-system operations, and *termination characters* are ASCII byte 0A. Instrument common commands are grouped into Auto Configure, System Data, Internal Operations, Synchronization, Macro, Parallel Poll, Trigger, Controller, and Stored Settings categories. Responses follow the same rules as commands and terminate by EOI, or end of message.

SCPI is a so-called "new language" that defines syntax and the structure of device-specific commands for instruments—especially those in automatic test equipment (ATE) systems. In other words, identical commands activate different instruments via software programming, which can also communicate via RS-232, as well as VXIbus and some other protocols and/or standards. However, it is essentially an IEEE-488.2 format extension and many instruments need SCPI complete command sets to complete a certain operation. Therefore, you will need some sort of identity listing for responsive communications. The IEEE-488.2 bus is a parallel bus.

Nominally, the GPIB is limited to 20-meter direct-cable connections, however, bus extenders and RS-232 serial links can virtually communicate with instruments located around the world since an RS-232 produces serial data for a telephone line and a device called an *autodialer*. For shorter distances of, say, 200 to 300 meters, fiberoptics cable is suitable. In our explanation, however, the RS-232 is actually the direct link that carries information between oscilloscopes, analyzers, computers, and a host of other instruments with RS-232 ports directly to ancillary devices, such as printers, plotters, or programmable hardware. Here, data rates are usually low and serial information from storage ROMs or RAMs are not necessarily extensive. Input and output commands usually do the programming with standard instructions common to general programming languages and systems.

VXIbus, based on both the GPIB and the VMEbus, extends the VMEbus specifically for instrument systems and will adopt SCPI as the instrument command standard. VXI was started as a consortium that now includes HP, Tek, NI, Wavetek, CDS, Racal-Dana, etc. This system enables multiple instruments to be placed on a single module or card with local lines to permit rapid communications.

Another group of buses are: Multibus II, NuBus, and Futurebus. The latter is said to become the next generation of VMEbus, and it is slated for use by the Navy in forthcoming developments. Backplane transceiver logic produces its high-transfer data rate.

As long as a bus is around, someone or something is usually traveling, but here we're primarily concerned with RS-232 and IEEE-488; further trips will be restricted to those two for now.

It's the hour now to begin working with National Instruments exclusively, until the end of the chapter. Much of the rest of the chapter is derived from lectures that were offered by National Instruments in the early 1990s, as are many of the line-cut illustrations and photographs.

Objectives

NI believes that data-acquisition systems for the 1990s consist of three prime elements: *acquisition*, *analysis*, and *presentation* (Fig. 8-3). These descriptions are rather large and nebulous; in simple terms, they mean using GPIB, VXI, and RS-232 instruments, as well as plug-in boards as sources for acquisition, and LabVIEW and LabWindows software packages for data analysis and presentation. Data presentation includes the display of strip charts, graphs, and graphical user interfaces (GUI) on both the computer screen and hardcopy output. In reality, rapid troubleshooting of any fairly noncomplex system needs little more than comparative scans for fault location if, indeed, the problem is at

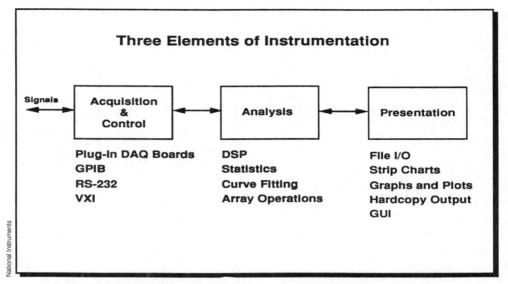

Fig. 8-3   *A generalized model of instrumentation systems.*

all obvious. The design software and output results must be evaluated quickly without reams of unnecessary data, thereby saving considerable time and temper. Remember, "tempus fugit?"

Because we've been describing and defining communications buses between and among various entities, let's sample several of these units and see how National Instruments makes them go. Thereafter, LabVIEW and LabWindows will be described to generally complete the overviews. The first is the GPIB interface for IBM PS/2.

An IBM interface

"Talker, listener, and passing controller" are included in all GPIB controller interface boards for IBM's PS/2 (Fig. 8-4).

Information to and from the micro channel are buffered for performance using a custom Turbo 488 chip. Outputs from this board go directly to the Address Latches and the Address Decode, but the Data Transceivers, Interface Logic, and programmable option select (POS) talk back as well, as does the DMA (Direct Memory Access) used for both input and output, and the interrupt logic. In turn, the latches, Data Transceivers, and the Bus Interface all converse with Turbo 488, a high-speed CMOS circuit that increases interface circuit performance, when used with the

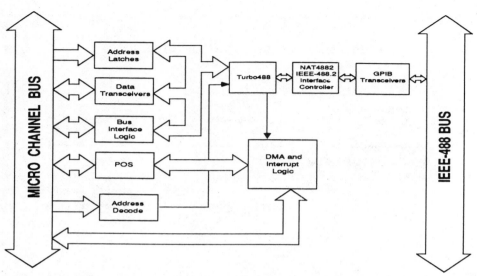

Fig. 8-4   *Talker, listener, and passing controls are all included in this IBM PS/2 block diagram.* National Instruments

GPIB TLC Talker/Listener, Controller IC. The Turbo 488 actually integrates logic from 16 ICs into one. Thereafter, signals pass through the GPIB Transceivers in 16-bit words, where byte-to-word circuits automatically provide byte-ordering in memory, and on the IEEE-488 bus.

The MC-GPIB is a very good example of hardware that exists between one equipment bus and that of the IEEE-488. Unfortunately, all this data and FIFO memory, handshake, synchronization, and routines isn't kindergarten simple—it's quite complex, but absolutely necessary. Your instrument or controller isn't going to use IEEE-488, RS-232, VXI, or anything else unless a proper interface makes the twain compatible . . . which should be somewhat evident by now. However, because you've already experienced a fairly rigorous routine, others should become apparent fairly quickly and with considerably more in-depth understanding. Therefore, let's try another example with somewhat different nomenclature, plus a different layout and application. This time, we'll travel with NI into audio, where A/D and D/A converters lurk and signal-to-noise ratios exceed 90 dB.

Hi-res audio I/O board for Mac II

This NB-A2100 (Fig. 8-5) possesses a pair of 16-bit resolution analog inputs, sampled at software-programmable kilohertz

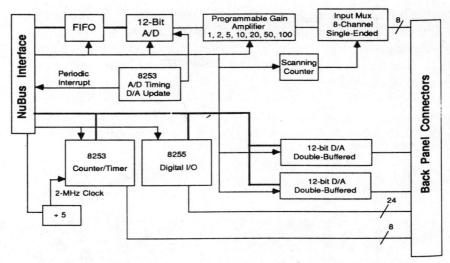

Fig. 8-5 *NB-A2100 does many things as the text indicates. It's for this Macintosh interface.* National Instruments

(kHz) rates of 22.05, 24, 32, 44.1 (for CD players), 32, and 48, respectively (the latter reserved for digital audio tape (DAT) recorders). This time, however, it's a NuBus Interface versus an RTSI Bus on the other end.

To identify, the NB-A2100 is a Macintosh II NuBus interface 32-bit slave board with a dual 32-sample FIFO (first in, first out) that precisely acquires digital data and generates at exact conversion speeds even though NuBus data might occur occasionally or irregularly (Fig. 8-3). The NuBus interface also has an 8K electrically programmable read-only memory (EPROM). NuBus, we might explain, was designed to optimize 32-bit data transfer and offer independent support for multiprocessors using a 10-MHz clock; it can transfer data at rates up to 37.5 megabytes/sec. With a typical data acquisition board, such as the NB-M10-16, the NuBus can transfer data between Macintosh memory and FIFO, a 12-bit analog-to-digital converter, a programmable gain amplifier, a scanning counter, and a pair of 12-bit D/A converters that couple to back-panel connectors. The NuBus Interface also outputs to a counter/timer and digital I/O as well as to a divide-by-5 2-MHz digital clock.

With the low-cost Lab-NBboard, analog input is received from a back-panel connector into an 8-channel CMOS multiplexer for the programmable gain amplifier, 12-bit A/D, and the FIFO output into the interface. There are 24 digital I/O lines in three 8-bit ports for input/output, bidirectional, or handshaking. Also, a pair of double-buffered 12-bit DACs serve two back-panel connections. Data-acquisition modes are either single-channel continuous or multichannel, with continuous scanning, and in either pretrigger or posttrigger conditions. A duo of 8253 16-bit independent counter/timers (three in each unit) compose the system controllers, and one counter can even increase the sampling interval whenever longer sampling times are required.

Combined with the NB-DSP23000 accelerator and LabVIEW software, the product of the NB-2100 board with the rest is illustrated in Fig. 8-6. Named the *Audio-Frequency Fourier Analyzer*, it is a very new audio instrument that was introduced in May,

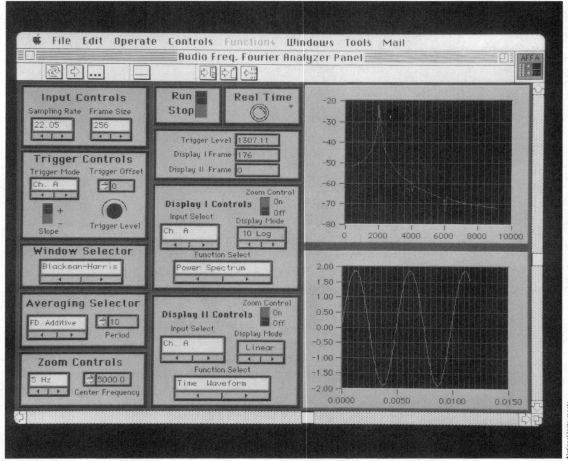

Fig. 8-6 *The very new audio frequency analyzer—a combination of NB-A2100, NB-DSP2300 accelerator, and LabView 2 software.*

1991. Fourier analyzers measure dynamic signals simultaneously, not individually, and are particularly useful below 100 kHz. Note: No signal analyzers do what spectrum analyzers do, but this technology is different. Both statistical and joint properties of two signals are possible, along with magnitudes and phases of all signal components. High accuracy is another of Fourier's attributes, which is especially desirable in critical, low-frequency investigations.

Software

Until now, major attention has been directed exclusively to hardware, its printout or display effects, and data-transfer between the various interconnect buses. However, complex analytical and testing hardware does not often operate independently of external direction. So, software must tell this equipment what to do and how to do it. Here, the programmer and his accumulated/stored commands and guidance information assumes a large share of design and troubleshooting operations.

Consequently, it was three years before NI President, Dr. James Truchard, and the Vice President of Research and Development, Jeffrey Kodosky managed to tame the "first, simple graphical way" to program and operate instruments on PCs. This, then, was LabVIEW, a visual graphical language for developing the virtual instruments.

Now, we can continue with NI-supplied descriptions of both LabVIEW and also LABWindows®, which also became available in 1991.

LabVIEW

The original LabVIEW software for Macintosh computers was introduced in 1986 as a means of solving programming and data acquisition problems for scientists, engineers, technicians, and "other nonprogrammers" to code and display their "wares" in an ordinary programming language. Graphically, you actually construct software models that become virtual instruments, such as image or likeness with wires and building blocks, or a front-panel interface with program controls, such as knobs, slides, and switches. Such blocks can operate interactively or be combined in successive orders in some larger or more extensive system. As an example, LabVIEW VI (using G) as a graphical programming language has node objects connected by wires to define programs as block diagrams. Nodes, then, become functional statements, and their connecting wires are variables; subVIs appear as subroutines.

LabVIEW II also has a compiler that converts block diagrams into machine code. Ready-made front-panel controls are selected and configured for a user interface (Fig. 8-7) where they are manipulated with a mouse-driven cursor and no I/O prompts. Next, block-diagram objects are selected and connected to observe data flow. The block diagram can be modified as needed and diagrams can be easily scanned for program structure, data, and whatever other complex operations can be obscured in a text-based program.

Execution orders are determined by data flow between objects for modular designs, construct layers of VIs, and access

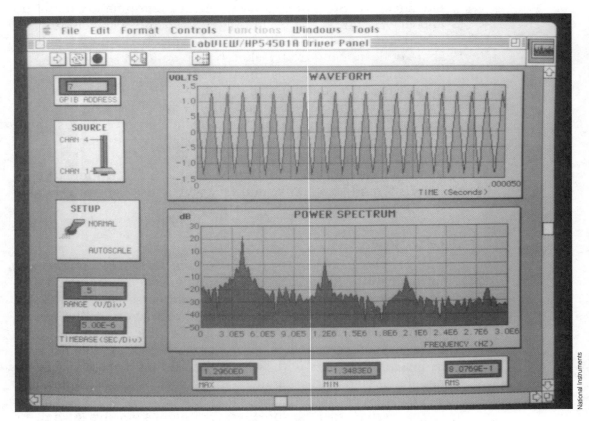

Fig. 8-7 *LabView is configured for user interface with controls and waveforms.*

subVIs for lower level operations. Best of all, custom panels can be operated by selecting controls and indicators from a menu, and LabVIEW contains hundreds of library routines for data-acquisition instrument control, storage, and instrument tasks. If you think conventional langauge programming is better than this equipment, external codes also be linked to a LabVIEW II block diagram.

The new LabVIEW run-time system permits LabVIEW users to use their VIs as an operate-only test station or as a repackage for a turnkey system. Because LabVIEW 2 compiles VI block diagrams into machine code (nothing stands alone), it is necessary to maintain the front-panel user interface. The end user, therefore, can work with front-panel controls and copy data to and from the front panel, but it can't modify the VI, so a source-code protect device is needed, too.

The foregoing should serve as a reasonable run-down on the LabVIEW software package and intent. If additional information is required, National Instruments has many newsletters and application notes available that will certainly be of aid in virtually any situation. Next, we'll proceed with LabWindows, which is a second software program.

LabWindows

For many who are already familiar with C or Basic, LabWindows emerges as an integrated software package that enables such regular programming methods with the needed additions that are required to build virtual instruments. Like LabVIEW, this software package provides data acquisition, control, analysis, and presentation at lower costs than before, with simplified modern equipment construction.

It provides integrated drivers for acquisition hardware, high-level drivers for specific instruments, auto-code generation front panels, analysis for specialty processing, and a new user interface library for graphical user interfaces. It will control GPIB, VXI, RS-232, and CAMAC instruments, and plug-in data-acquisition boards in both production test surroundings, and R & D laboratories. LabWindows also has a special interface, known as a *function panel* for each of the functions in the libraries, that programmers can use to develop functions while generating a source code for certain applications. Each panel offers on-line help for operation, and there are over 120 high-level drivers for the various buses and their instruments. There's no requirement to know low-level programming here.

The User Interface Library has a graphical editor and various functions for graphical programs and applications. The Analysis Library permits array and matrix manipulation, complex arithmetic and statistics (such as histograms and mean standard deviation). An optional Advanced Library also offers Fast-Fourier Transforms, curve-fitting, digital filtering, windowing, real and complex vector and scalar qualities, as well as high-level statistical analysis.

According to Michael Santori of National Instruments, LabWindows has four work areas: the programmer to edit and run complete programs, the interactive window to process code sections out of context, an input/output window for screen output, and an error-list window for the tabular viewing of program syntax errors. The software runs "under" MS-DOS; therefore, it accommodates a considerable number of the better instruments and, being compatible with QuickBASIC and C, interactive programs can be compiled and run on any machine in a given application, including "floppy-disk only" computers.

Heretofore, LabWindows has been limited by a 640-kilobyte-memory restriction of DOS. With the DOS/16M DOS Extender, users are able to work programs up to 16 megabytes of memory—even if only two megabytes of memory are installed in the computer.

According to National Instruments, LabWindows is "simple." Many programming aids are in the system, a number of debugging assists are included along with extended memory, and special codes are generated with function panels as interfaces for executing library functions without typing and editing by selecting pictured controls on the front panel. Graphical user interfaces can be created to combine graphic panels and menus for simple interactive operations. Program users can see measurement data and control system operations directly from their interfaces.

They can also select full-color controls, graphs, and strip charts. Also, custom panels can be created "by importing PCX images."

That's about the extent of information that is generally available on software. National Instruments has an excellent 1991 catalogue of several hundred pages devoted to explanations and characteristics of its highly useful products filled with both software and hardware applications and illustrations. For all who need this sort of information, you'll find it invaluable.

NI was deliberately selected as a forerunner in these special disciplines and it will certainly occupy an important position in the laboratory now and in the future.

What your author foresees will become "virtual" test instruments generated and applied through these same types of software (automated or otherwise), appearing in many or most of the larger laboratories and service centers around this country and abroad are doing is a much more comprehensive and accurate assignment of answering outstanding questions than did the older-type of operator-adjusted and calibrated (or not) equipment. You'll certainly enjoy self-calibration, close-tolerance accuracies, a minimum of stand-alone gear, excellent large-screen software projections of one form or another, and a library of disks that are double and triple the number of what some have today.

One would also expect that such buses as GPIB, EIA-RS 232, and others, as well as their interfaces could be assigned as punch-key software, simply awaiting a programmer's beck and call. If this comes to pass, and layout or troubleshooting assignments further simplifies, a host of such LabWindows and LabVIEW will be evident in laboratories all around the world. We happily look forward to these developments in the same mode and manner that NASA exhibited in the 1960s when RCTL gold-bonded flatpacks made possible our unmanned satellites between here and the moon!

Futurebus +

A separate topic that we haven't touched on is *Futurebus* + . Yes, this is another talk and answer arrangement because it has no connection with the other National Instrument endeavors and aspirations at this time. The expectation is that Futurebus, with IEEE approval, can and will become a player in data transmissions because many manufacturers (such as Motorola, Unisys, DEC, Sun Microsystems and others) are actively involved. It's allegedly immune to changing technologies, has global addresses, special module communications, and parallel protocols for data exchange, plus control and status registers for bus management.

Said to be an open bus standard, it will accommodate a host of microprocessors and other significant subsystems without obsolescence. Its "arbitration" process is in parallel with bus data transfers, and individual modules have a unique 8-bit number during arbitration, with one becoming predominant when it possesses the largest arbitration number. Each number has a "round-robin" bit and a 5-bit special identifier.

Basically, Futurebus offers a 64-bit data and address bus, a 32-bit address/data subset, and a 128- or 256-bit data bit stream. Data transfers are entirely asynchronous and are not locked to a specific clock rate or frequency. Information is first placed on the bus, and all is transferred in three phases: beginning with connection (to a particular slave); optional data transfer; and a disconnect phase, which both ends the transaction and shuts off the slave.

In motion, transactions generate bus beats for master/slave synchronization control and address information, and the slave waits for information lines to settle, detects the sync symbol(s), then stands by for its own receivers' skew. Afterwards, slave information captures and engages specific transmitter lines. This process ends in an electronic handshake.

Hardware configurations A and B are already committee-approved, while C and D might still reside in limbo until all are in agreement. Both A and B can handle 32- and 64-bit addressing. The difference is that A requires backplanes to support 64 bits with default to 32 bits; B supports a 128-bit path, defaulting to 64 bits, then to 32 bits, in that order.

Information is available, but additional changes might conflict with our initial research. Therefore, further description might even hinder, rather than enlighten, so we'd better stop here.

❖ 9

Signature analysis

PROBABLY NOT THE "IN" TOPIC FOR THE 1990s IN THE BEGINNING, but with the spread of digital instrumentation and its applications, you might have use for signature analysis sooner than you think. Therefore, its inclusion in the final chapter should seem appropriate to some and a welcome addition to others. Logic analyzers, themselves, have changed with the advent of advanced technology and wholly new test sets—some of these are small enough to "tote" around in a light shopping bag and are priced quite reasonably, considering comparative costs of other fairly sophisticated logic and DSO equipment. In this case, we'll use a new unit by Hewlett Packard for our reference and try to give you a working idea of what it will do. In addition, we'll explain a few logical numbers that are relevant to the subject. Before the chapter's end, you might just find something that you can use or consider as new software changes the format of many signals and systems. One of these could be yours.

Logical numbers

But, perhaps, before venturing directly into this test instrumentation, we should say a word or two about digital logic and the words and symbols that go with it. For many who are already familiar with these basic topics, skipping this particular section will do no harm, but we might introduce an idea or two that might prove informative. So, continue at your pleasure.

For our purposes, decimal, binary, hexadecimal, ASCII, and EBCDIC are logical numbers. We won't tax either your patience or sensitive powers of perception with long lists of number conversions or tedious explanations of what you should already know, but we will make some comparisons of different families that could both remind you and make such number-language conversions easier to recognize and execute. For any who have no logic instruction, you'll need a great deal more schooling in this discipline than we plan to present here. A good foundation in the 1's and 0's business amounts to much more than a few basic pages of oscilloscope book text. We can't possibly offer them all of this information now, especially because most logic equipment users have intensive programming on the subject already drilled into them. Consequently, we'll proceed on the assumption that most readers will have a working familiarity with many of the systems that are superficially covered, and we can rather easily pick up the rest.

Decimal radix

Because most humans learn to first count on their fingers (before using a little electronic calculator), the decimal system was a convenient method of showing their skills. This system became known as the *radix* or the base of the decimal system and it becomes a power of 10. Therefore, if you wish to write a number such as 2 345, divide such numbers into segments, and let each segment represent a successive power of 10. Start, however, from right to left, with divisions numbered successively as 1's, 10's, 100's, and 1 000's, as follows:

$$2 \ (\times \ 1 \ 000) \ 3 \ (\times \ 100) \ 4 \ (\times \ 10) \ 5 \ (\times \ 1)$$

In multiplying and then adding all these numbers together, you have an answer of:

$$
\begin{array}{r}
2\ 000 \\
300 \\
40 \\
\underline{5} \\
\end{array}
$$

or　2 345　which is precisely what we wanted.

The powers of 10 keep repeating an extra zero after the initial digit, 1. You can therefore see that 10^0 is 1, 10^1 is 10, and 10^2 is 100, etc. Hence, each time some number goes from two to three digits, for instance, it becomes hundreds. If there are four digits, you can read it in the thousands because three digits following a fourth means 10^3, or three "zeroes." So, this number automatically becomes 2 345.

Common logarithms, of course, are all keyed to the base 10. You will find that the antilog of 0 becomes 1, that of 1 is 10, and that the antilog of 2 equals 100. Reverse this little procedure and:

$$\log_{10} 1 = 0, \log_{10} 10 = 1, \log_{10} 100 = 2, \log_{10} 1\,000 = 3$$

Similarly, $10^0 = 1$, $10^1 = 10$, $10^2 = 100$, etc., just as we showed before. However, if we arbitrarily change base 10 to base 2, the answers would become:

$$2^0 = 1, 2^2 = 4, 2^3 = 8, \text{ and } 2^4 = 16.$$

You might write the final number, as an example: $\log_2 16 = 4$. Quite suddenly, but simply, we have translated from the decimal system to one of binary (base 2), which digital logic accepts with relish.

Binary

Binary, however, appears in a little different form because, like the microprocessor or computer, its information consists of purely 1's and 0's. When some digital device is programmed to accept and operate with such language, then all sorts of results are possible. Executing powers of 2, for example, we can show in binary that a number such as 525 can be developed as follows:

$$
\begin{aligned}
2^9 &= 512 \\
2^3 &= 8 \\
2^2 &= 4 \\
2^0 &= 1 \\
\hline
& 525
\end{aligned}
$$

Observe that 2^1, 2^4, 2^5, 2^6, 2^7, and 2^8 are not used because they all represent zeroes. Written in pure binary, the digits, arranged in groups of four look like this:

512	0	13
0010	0000	1101

This all equals 525, because we simply add the results as before when using powers of 2. The group of four have been arranged accordingly:

Decimal   8 4 2 1
Binary   1 1 1 1

expanding, as another power of 2 in succession.

Zeroes, as you can see in the preceding example, don't count. Therefore, any initial full group of four (with all 1's) represents a maximum number of 15; that is, 0 through 15 since 2^0 equals 1. This count then makes the 9th power of 2 fall where it is in the 3rd group of four, because each power is counted in ascending order from right to left. As you can see, unless one is directly involved in this sort of logic, working with it occasionally becomes quite difficult and errors are often commonplace.

Binary coded decimal

To try and relieve this condition, binary coded decimal (BCD) was constructed, but it soon was discarded because it had no base 2, 10, etc., was limited to digits between 0 and 9, and evolved nothing more than a pseudosimplistic method of working with binary numbers. True, every group of four bits represents digits 1 through 8, but it never changes thereafter. Therefore, our same 525 number in decimal would be expressed as follows in BCD:

Decimal   5   2   5
BCD   0101 0010 0101

As you can see, anyone could learn this very simple 1, 2, 4, 8 routine in a matter of minutes. However, many more intricate rules were added to the exercise than we have space to present, so it wasn't popular. The machines didn't like it either. Fortunately, other systems quickly became more agreeable.

Octal

Octal, meaning 8, answers to $\log_8$, and it becomes another "anchored" number system with creditable antecedents. First, go back to exponent notation and translate into logarithms:

Exponential	**Logs**
$8^2 = 64$	$\log_8 64 = 2$
$8^4 = 4\,096$	$\log_8 4\,096 = 4$

$$8^6 = 262\ 144 \qquad \log_8 262\ 144 = 6$$

This system isn't difficult (if you have a good calculator), and it does, indeed, expand the ability of numbers to cover much greater territories in fewer steps. As a result, each position becomes a power of 8. If you wanted to work 525 between octal and decimal number systems, the result would be:

Octal becomes:

$$
\begin{array}{ccc}
64 & 8 & 1 \\
\times\,5 & \times\,2 & \times\,5 \\
\hline
320 & +\,16 & +\,5 \quad \text{or 341 decimal}
\end{array}
$$

Refer back to the previous example if you wish to compare octal with binary or BCD. Notice that the decimal figure is always less than the octal.

Hexadecimal

The name sounds ferocious, but it really isn't. It's simply more of the same but, this time, to the base 16 with a few extra twists to keep life interesting. Here is another code broken down in terms of 4-bit binary, which is read normally between 1 and 8. However, we also have letters A through F tacked on to represent decimals from 10 to 15. As usual:

$$16^2 = 265 \text{ and } \log_{16} 256 = 2$$
$$16^3 = 4\ 096 \text{ and } \log_{16} 4\ 096 = 3$$
$$16^4 = 65\ 536 \text{ and } \log_{16} 65\ 536 = 4$$

What's really happening is that people think in terms of decimal (base 10), but computers and such respond to binary (base 2). Hexadecimal really represents a compromise so that any number between 0 and 15 can be shown as 16 separate bits and each hex digit can be found in any four binary bits. Once more, we'll exhume our usual number 525 (in decimals) and see how it translates into Hex:

$$
\begin{array}{ccc}
256 & 16 & 1 \\
\times\,5 & \times\,2 & \times\,5 \\
\hline
1 & & \\
280 & +\,32 & +\,5 \quad \text{or 1 317 Hexadecimal}
\end{array}
$$

Now, let's translate this Hex number into binary using 1 through 8 and one of the A to F letters standing for 10 through 15. Our number, of course remains 1317.

D	1	7	Hex
1101	0001	0111	Binary

That looks pretty easy, doesn't it? Now, both 1's and 0's equipment (so-called machines) and their human operators get along just dandy. The missing link has been discovered—just like the IEEE bus, as you will see later.

However, we haven't run out of useful codes yet, and two more need just a little discussion and a pair of routine examples to make life just a little easier before we present the somewhat difficult equipment characterizations and operations.

ASCII

ASCII is commonly called a code and it represents the *American Standard Code for Information Interchange.* As you should quickly recognize, any set of coded symbols that is reducible to binary can transmit data to a binary machine. Of course, ASCII fits this definition. If your equipment, for instance, has an ASCII keyboard then it "sees" an ASCII-coded symbol each instant that you strike a key and operates on that basis.

This time, however, instead of doing the usual number routine, we'll borrow Tektronix' printed ASCII Code Chart (Fig. 9-1) that shows the entire routine in binary, hex, and ASCII. In this way, you can visualize the complete three-way relationship and be guided accordingly.

EBCDIC

EBCDIC, another code equipment language, is illustrated in Fig. 9-2. Translated, EBCDIC represents *Expanded Binary Coded Decimal Interchange Code*, and it is used by the largest computer manufacturer in the world on their machines. So, this one can't be ignored either. As we progress with equipment applications, you'll see that all languages and codes have their special places in the very large and growing world of machine language applications.

To use these many codes and languages in the various transmission mediums, such data must be moved in a practical bit stream. The next section deals with such methods directly, but briefly.

HIGH \ LOW	0 0000	1 0001	2 0010	3 0011	4 0100	5 0101	6 0110	7 0111	8 1000	9 1001	A 1010	B 1011	C 1100	D 1101	E 1110	F 1111
0 0000	NUL	SOH	STX	ETX	EOT	ENQ	ACK	BEL	BS	HT	LF	VT	FF	CR	SO	SI
1 0001	DLE	DC1	DC2	DC3	DC4	NAK	SYN	ETB	CAN	EM	SUB	ESC	FS	GS	RS	US
2 0010	SP	!	"	#	$	%	&	'	(	)	*	+	,	–	.	/
3 0011	0	1	2	3	4	5	6	7	8	9	:	;	∨	=	∧	?
4 0100	@	A	B	C	D	E	F	G	H	I	J	K	L	M	N	O
5 0101	P	Q	R	S	T	U	V	W	X	Y	Z	[	\	]	<	_
6 0110	`	a	b	c	d	e	f	g	h	i	j	k	l	m	n	o
7 0111	p	q	r	s	t	u	v	w	x	y	z	{			}	~	DEL

BINARY — HEX — ASCII

Fig. 9-1   *The American Standard Code for Information Interchange transmits data to keyboard binary machines.*

Tektronix

HIGH \ LOW	F 1111	E 1110	D 1101	C 1100	B 1011	A 1010	9 1001	8 1000	7 0111	6 0110	5 0101	4 0100	3 0011	2 0010	1 0001	0 0000
0 0000	SI	SC	CR	FF	VT	SMM	RLF		DEL	LC	HT	PF	ETX	STX	SOH	NUL
1 0001	IUS	IRS	IGS	ITS		CC	EM	CAN	IL	BS	NL	RES	DC3	DC2	DC1	DLE
2 0010	BEL	ACK	ENR			3M			ESC/PRE	EOB/ETB	LF	BYP		FS	SOS	DS
3 0011	SUB		NAK	DC4					EOT	UC	RS	PN		SYN		
4 0100	|	+	(	<	.	¢										SP
5 0101	¬	;	)	*	$	!										&
6 0110	?	>	_	%	,	¦									/	-
7 0111	"	=	'	@	#	:	`									

BINARY	HEX										
1000	8		a	b	c	d	e	f	g	h	i
1001	9		j	k	l	m	n	o	p	q	r
1010	A		~	s	t	u	v	w	x	y	z
1011	B										
1100	C	{	A	B	C	D	E	F	G	H	I
1101	D	}	J	K	L	M	N	O	P	Q	R
1110	E	\		S	T	U	V	W	X	Y	Z
1111	F	0	1	2	3	4	5	6	7	8	9

BINARY — HEX — EBCDIC

Fig. 9-2 EBCDIC, a prominent computer code language used by the world's largest computer manufacturer.   Tektronix

Data transfer

Two prime types of data transfer are used between equipment (machines) and the outside world. One way is parallel, and the other way serial. Each has its particular application, depending on whether you wish to offer a great deal of information in parallel form over a short distance, or to transmit and receive yards or miles down the pike. In parallel, of course, many leads or lines are arranged beside one another so that maximum intelligence becomes available as two pieces of equipment interact (Fig. 9-3). For instance, 5- or 10-bit multiple transmissions would require 5 or 10 lines. In ASCII, one bit of a 7-bit character line is known as *parallel-by-bit serial-by-character*. Naturally, other types exist, but for short hauls, parallel transmissions offer a much quicker means of transferring 1's and 0's from one machine to another.

Serial information interchange is a considerably more common means of communications, because it has been in use from the era of the telegraph and the venerable Morse code. Even in

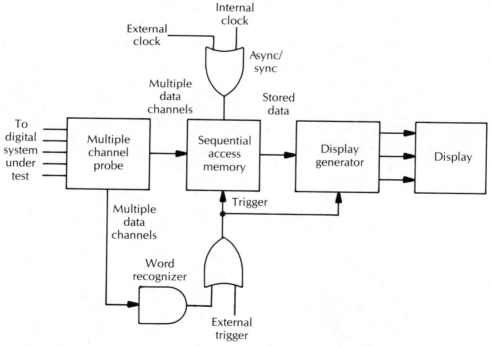

Fig. 9-3 *Word recognizer triggers any logic analyzer when parallel words match whatever is selected. Pre-, center-, and post-trigger modes are also available.* Tektronix

those days, transmissions were in bit streams at a good baud rate, if operators could muster a steady "fist." Now, of course, memory storage, extremely fast ICs, and superior means of cabling or microwaves, permit megabit data transfers over almost any required distance with considerable accuracy.

Sync

Whether synchronous or asynchronous, some definite sync signal must exist between transceiving equipment. In other words, just like marriage, one partner has to know what the other is doing. An out-of-sync condition is wholly unsatisfactory.

Consequently, the receiver must receive some bit sync from the transmitter so that it can stay in lock with starts and stops of the data stream. Without synchronization, transmitted information would be totally lost and the product would be incomprehensible garbage. Obviously, the receiver deserves better.

So, there must be both bit sync as well as a character sync in order for the two pieces of equipment to stay locked. In *serial synchronous transmissions*, information is sent and received in blocks, according to defined start and stop references. They can be of any assigned bit length and all are transmitted sequentially, usually at considerable speed and quantity. *Serial asynchronous transmissions*, on the other hand, transmit information at some random rate (wherever convenient, for instance); words or characters can be of any preassigned length. There are both start and stop bits, however, that frame transmissions, and the start bit warns a receiver that the transmission is commencing and to begin sampling for incoming characters.

Modulation/demodulation

Some people lump modulation and demodulation functions into a single piece of equipment, called a *modem*. It can convert a digital signal into analog for certain data transmission, then convert this analog information back to digital for use, possibly by another computer. If you are specifying a system, however, and you require only a modulator or demodulator, *don't* specify a modem because it could cost you more than double either of the other two systems by themselves. You'll find A/D and D/A conversions becoming much more in vogue as the 1990s progress because of the enormous linkage between all sorts of microwave, satellite, and fiberoptics high-speed, large-bandwidth carriers.

In actual modulation/demodulation techniques, modems use amplitude modulation (AM), frequency modulation (FM), or phase modulation (PM). In AM, the carrier frequency is modulated via its envelope, and its amplitude then predicts either a 1 or 0. FM, of course, deviates a constant carrier, and digital samples these deviations and places a 1 or 0 for frequency decreases or increases. PM looks at voltages of similar amplitude and frequency, but phase changes within the frequencies constitute all modulation. This phase change during specific times is then represented by a 1 or a 0 as the information changes.

Modem makers have combined all three methods, in some instances, and produced di-bit or tri-bit combinations in certain arrangements, with each becoming a level for amplitude, deviation, or phase of the processed intelligence. Machine talk then takes the form of simplex, half duplex, or full duplex. So, the machines can have one-way transmission, one-way at a time transmission with full bandwidth and maximum data rates, and transmissions/receptions in both directions at the same time, respectively.

A digital circuit tester

Hewlett-Packard calls their HP 5006A signature analyzer "a digital troubleshooting tool" and it most certainly is. Function and polarity switches and a 4-digit readout of characters (0 through 9 and ACFHPU, with a probability of 100% in detecting single-bit errors and 99.998% chance of finding multiple-bit errors) are on the front. With internal logic compression, bit streams are compacted into 16 bits so that only signatures need be compared.

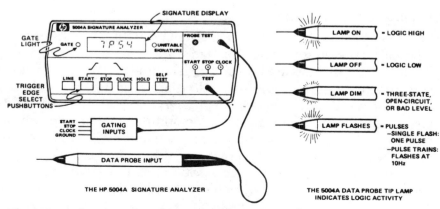

Fig. 9-4 *A drawing of a 5000-series signature analyzer.* Hewlett-Packard

These are "cyclic redundancy codes" that check the various blocks of information whether by a specifically generated test pattern or via ancillary inputs from other types of information, where logical bit streams have documented values and frequencies. Thereafter, signatures are stored in memory once the probe switch is engaged. Because the final 32 signatures are stored, the signatures can now be compared in groups as composites, rather than as signatures that follow individual probes simply by Recall.

Signature characters and operation

Sixteen characters represent the readouts of the analyzer's operational display, beginning with O and ending with U (Fig. 9-5), but only 7 can appear at any one time. Although all are represented by the usual 0000-to-1111 binary groupings, this is no real hexadecimal conversion but was designed for easier troubleshooting and faster alpha/numeric recognition.

As in all binary logic, 1s and 0s represent *highs* and *lows* (or *ons* and *offs*), as switching or toggling proceeds during equipment operations. Here, when a certain node or circuit is probed and a high appears, the analyzer's gate opens internally. The gate has three controls: start, stop, and clock. So, a series of logic or bit streams can pass through for a certain period before being stopped; clocking can occur on either the rising or falling edges of the input signal—all of which is determined by the analyzer's operator.

The clock synchronizes the analyzer and the DUT, logic levels are clock-sampled, and the resulting data information is compressed into a signature. Figure 9-6 is an excellent illustration of

DIGIT	DISPLAY
0000	0
0001	1
0010	2
0011	3
0100	4
0101	5
0110	6
0111	7
1000	8
1001	9
1010	A
1011	C
1100	F
1101	H
1110	P
1111	U

Fig. 9-5 *Characters and binary digits display and store signature readouts.* Hewlett-Packard

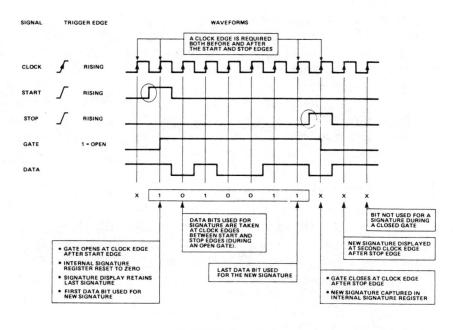

Basic Gate Operation

Fig. 9-6 *Clock and data operations within an analyzer.* Hewlett-Packard

all these actions; binary highs and lows are tallied and read out thereafter in signature format. Observe that the clock starts and stops on the rising edges of the first and second pulses, which are shown beneath the clock. A falling edge could be selected also, if that was the deliberate intent, and gate timing is also a convenience decided by circumstances and the controller.

In some situations, the gate will neither open nor close. This lockup can occur if no clock edge is apparent before and after the start-stop edges or if the start and stop connect to the same signal, but trigger on opposite edges. Then, the absence of a clock edge for low-level detection on a logic low won't open or close the gate either.

Occasionally, the start/stop trigger mechanism is too short, delayed, or out-of-phase (frequency) with the analyzer's clock. The gate will not open if the timing is slower, but could, conceivably operate with a higher frequency clock. Also, the clock sometimes operates following the intended start edge. Occasionally, the stop edge goes undetected, as a result of the counter's clock not working.

The prior explanation sounds simple enough, but the instrument itself isn't. It has logic gates, read-only memories, decode memories, the clock, control and data buses, and a Z80 central-processor unit (CPU) (Fig. 9-7). Start/stop enables it to connect to a bit latch as an I/O port, where start energizes on the input's rising edge and stop on a trailing edge. All ROM 2 addresses are read on the data bus, which results in a signature. Once the program locations have been fully disclosed, the gate is closed and another byte or so of signature topping begins. ROM 1 does much the same, except that it stores a ROM 2 stimulus program and digital address counterparts from 0000 to 07FF. Signatures are then combined from the two ROMs and read out accordingly, across the appropriate bus. Remember that Clock, Gate, and Start/Stop commands must all coordinate so that the gate will open and close when desired and that enough samples are taken to fully decode the all-important signature.

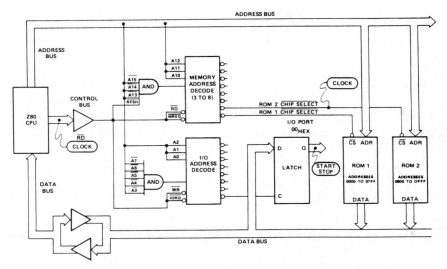

Fig. 9-7 *A block diagram of a similar analyzer with its microprocessor, memories, latches, buses, and ROMs.* Hewlett-Packard

Once a signature is known, however, it can be used as an all-time reference if you're measuring under similar conditions. If not, nodes at other PC-board sections should alert both you and the analyzer to problems that have arisen or still exist. If you're starting out ''cold'' on a defective PC, special probing and a very keen nose for certain ''alphabet soup'' could lead toward any useful system or subsystem signature. Otherwise, pump in a signal, engage the trusty old oscilloscope and work a couple of channels until something rational appears. Otherwise, the manufacturer might have the information you sorely need, and they might even xerox or fax a copy to your doorstep. It's been our practice in the past, however, to examine power supplies first if any voltages seem either puny or absent altogether. All really depends on the equipment under test and the method that has been selected for testing.

For automobiles, games, some television circuits, and even computers, signature analysis does seem to be at least one solid means of finding faults with minimum effort. The days of troubleshooting digital, or even analog equipment, for that matter, with either digital or analog voltmeters should be long past. We don't make a habit of "putting down" meters, but some sort of analysis is needed today, rather than the tired old procedure of hunt and pick. Why not develop pertinent signatures for some of your smaller, easily managed units, and try a few signatures for posterity. I bet you'll be happy that you did!

Accuracy

These figures might surprise you. Hewlett-Packard says that single-bit errors are 100% detected, provided that the defective bit is already in the analyzer. Multiple errors collected by the analyzer have a detection probability of 99.998%. Close enough for government work?

By the way, it isn't necessary to keep the analyzer's gate open for extended periods. The absolute accuracy is for 16 or less data bits and the others are within 0.002%. Therefore, a bit stream at any node has a probable accuracy within 2×10^{-3}. We hope that you're impressed, because these figures are one of the overwhelming reasons that chapter 9 was included in this book. The accuracy of a good signature analyzer is really remarkable.

Are the same signatures available at the various nodes without fail? If the equipment under test is receiving the identical inputs, there's no reason for change unless the system has developed problems. If you control the gate openings properly, you should see the same results every time.

The gating can be controlled a couple of ways. Start and stop connect to separate (different) signals. Start and stop can trigger on either rising or falling edges. These conditions are illustrated in Fig. 9-8, which shows rising and falling edge triggers, as the gates open and close.

However, start and stop triggers can operate on an identical signal, both on the rising or falling edges, but the gate opens for the necessary bit-stream time, then closes to complete the cycle, as programmed. You see, therefore, that many ways are available for flexible use of your analyzer in all varieties of circumstances.

Are signals ever unstable? Absolutely! The gate has to be open for identical edge times and logic levels, plus timing from one gate opening to the next. An Unstable Signature light announces this condition, and it glows for 100 ms when signals in any gate cycle aren't the same as those from the previous gate cycle. Obviously, some signatures are repeatable and nonrepeatable. Identical nodes should be repeatable, but if the signature each time from the same node is different, indeed, it is a serious problem. A possible cause could be that of RAM data, as opposed to information in the ROM. When systems are shut down and then revived, the free-running microprocessor has no means of storing a pattern in the RAM, because it is a random-access memory. Therefore, this RAM and other noninitialized devices can't be on the data bus, and the ROM memory will be measured. Literally, adjacent signatures are equal for all measurements, but unequal signatures are unstable because they're not the same.

To be sure that equal signals are repeatable, first turn the DUT off and then on; reset the analyzer, and ground or V_{cc} the probe before the next measurement; then start your signatures.

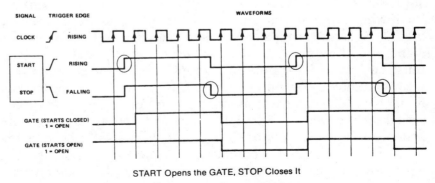

Fig. 9-8 *Predictable gating for openings and closings.* Hewlett-Packard

Noise

Noise is always a disturbing factor in any measurement, whatever the cause or the instrument. Extraordinary noise can even obscure a weak signal; ordinary noise simply causes circuit and system problems that often produce false-test readouts. In analog measurements, signal-to-noise (S/N) is one of our most critical parameters. Digital readouts can be affected also, but not in exactly the same way, nor to an equivalent extent. So many commercial and consumer products are going digital to avoid a great deal of noise and to increase information accuracies.

Asynchronous noise can originate anywhere within the system, especially if the digital or analog signal lines have little shielding or routing. Much of this noise has already been eliminated by computer-aided design (CAD) work stations, where complete systems are synthesized, studied, and largely computer-corrected before being set in hardware.

Synchronous noise usually arises from the system clock or its clocked subsystems, especially while undergoing high-to-low or low-to-high transitions. Often termed *glitches* or *aberrations*, the tester will pay them little attention, unless they appear at the clock switching edge of the analyzer. Synchronous input noise and stop/start noise is largely ignored by the analyzer. However, asynchronous noise on stop, start, and data on the clock edge can result in false, nonstable, or nonrepeatable signatures. Be on your guard here. A signature analyzer takes some "getting used to" before its full utility is apparent. After a few trials and errors, most of its measurement benefits should become second nature.

In the event that noise sources are not readily apparent, be sure that the tester is securely "grounded" to the DUT. Look out for further clock edges that occur when the analyzer's gate is closed, because startups can cause instability and nonrepeatable signatures. Common coaxial cabling between tester and testee must be fastened securely to prevent circulating currents from forming, known conventionally as *ground loops*. All probe commons should be as short as possible to negate any undesirable signal pickup. This is especially true in television and fast digital systems, where intermodulation and/or crosstalk can become real problems. Further, in a computer or a multidigital-instrument environment, power lines and inefficiently shielded adjacent equipment always offers potential interference to both the DUT and your own analyzer.

Simple filtering and even surge-suppression networks are often ineffective, especially if offered by inexperienced or "bargain" manufacturers. Plan metal-oxide varistor semiconductors (MOVs) are nice, but usually they require additional support and fast time constants to soothe the savage sea of overshoots, preshoots, voltage spikes, and whatnots. Your better equipment needs substantial protection over and above that which is usually built-in, particularly in an obviously hostile environment. Lastly, common (ground) connections should always be made to a single point on the DUT chassis, if at all possible. This procedure usually avoids any extraneous modulation effect on DUT electronics and your signature readouts. The single exception occurs when more than one common is designed into equipment. Be sure, therefore, that this is not the case when troubleshooting a DUT that's unfamiliar. In solid-state electronics, "familiarity" never breeds contempt.

Manual troubleshooting

In computer-assisted systems, signatures usually are compared automatically. In small equipment, where digital circuits, buses, and/or microprocessors can be investigated, discrete probing in suspected areas usually brings rapid results. Pick a few nodes from an appropriate schematic, check their (hopefully) prerecorded signatures for immediate faults.

If a composite signal and a node test point are available, proceed to this one first before tracing backward to the fault origin. Thereafter, you can "divide and conquer" in whatever direction the problem takes. This hunt should proceed with individual signature checks to the subsystem that is apparently defective. You can also use the Recall signature mode in the analyzer that remembers the last 32 signatures probed. You should also know that address and data buses do have composite signatures, which should make them easy to verify. Recalled signatures, of course, appear in reverse order to those originally stored. However, signatures that were measured more than once retain only the last probing, rather than any previously stored. In other words, old signatures are automatically erased and written over as new probings occur. *Composite signatures* are always important, because they are the actual sums of the individual signatures.

If you wish, these can be summed alphanumerically by first converting signature to hexidecimal for the changeover, and then back to freerun for the final answer. Remember that Hex has common address numerals, just like the signatures.

$$0, 1, 2, 3, 4, 5, 6, 7, 8, 9 \ldots 15$$

but the letters differ, which are:

A B C D E F	**Hex**
A C F H P U	**Signature**

So, if you would convert the following:

Signature	Hex	Address
A86P	A86E	0
CCHU	BBDH	1
3AFP	3ACE	2

0127

letters, of course, represent numerals 10 through 15

Carries in excess of 16 bits (2^4) are discarded because signatures have only 4-way readouts.

This equivalency is only several of the possible 16 symbolic sets of four letters/numbers that you might assign. So, 0, 1, and 2 would be applicable here, but you would have to designate address number 7, etc. to generate the complete reconversion to Signature. Bluntly, it's much easier for the HP 5006A to do the job. *Mistooks* are easy in this business! Notice that you're really substituting letters, not numbers. Addresses, naturally, remain the same in whatever slot that's designated.

Undoubtedly, there's more to these signature applications than offered here. However, with a reasonable introduction to this engineering art of node and system troubleshooting, we should have helped convert some occasional disbelievers. Those who don't see at least some advantages to this procedure, might have avoided an interesting opportunity.

❖ 10
Vectorscopes and vectors

ORDINARILY, YOU MIGHT PAY LITTLE ATTENTION TO RELATIVELY unfamiliar equipment, such as vectorscopes, whether NTSC or gated rainbow generators. However, in today's video age, the vectorscope is fast becoming one of the most important tools in color analysis. Although phase change does not necessarily require signal frequency change, chroma phase shift produces a miserable video image, which neither industrial nor consumer viewers will tolerate. As monitors and receivers improve even more, proper chroma oscillator/demodulator adjustments will become as much of a sacred ritual as automobile ignition timing. True, pings and phase don't precisely rhyme in English prose, but engine knock on the road and purple people at home are somewhat akin—especially after a bad tuneup. The family car, like the family television, are both precious commodities, worth their respective weights in a peck or bushel of still popular U.S. greenbacks.

What is a *vectorscope*? Tektronix calls it an oscilloscope with a circular time base used in chroma signal examinations. Actually, it produces a polar plot of chroma phase and amplitude, its radius represents chroma amplitude and its phase angle that of chroma tint or hue.

Such an NTSC (National Television Systems Committee) vectorscope (Fig. 10-1) is able to measure luminance amplitude, chroma phase and amplitude, differential phase, and differential gain. Additionally, time delay between signals can also be checked, as well as studio camera phase differences, which might also include connecting cable lengths. Fancy vectorscopes can even show dual displays, including time-base measurements.

Tektronix

Fig. 10-1 *Typical NTSC vectorscope response used for chroma phase and amplitude measurements.*

Vectorscopes of the NTSC variety are highly useful in studio-transmitter, camera and system setups, color renditions, transmitter levels, and chroma outputs. However, video products that have variable brightness, hue, and chroma customer controls offer a problem because their levels are not arbitrarily fixed. Consequently, color receivers aren't the best subjects for NTSC waveform generators and their inquiring vectorscopes. Nor are any video units that don't offer either I and Q or R-Y and B-Y outputs. Although the same holds true for gated rainbow generators, their 10 color bars combined with highly useful vector petals, are heartily recommended for consumer products because variable levels can be adjusted to produce the required vector. In addition, no special vector graticule is required with rainbow generators, because the pattern (if satisfactory) has the same shape everywhere, and all desirable characteristics can easily be delineated on an ordinary oscilloscope's 8 × 10 grid, anyway. As you will shortly see, the perfectly phased vector has just 10 symmetrical voltage petals that appear just 30° apart.

The fundamental difference between NTSC and "sidelock" rainbow generators and displays, besides having considerably more electronics, is their time base. As you might expect, the NTSC variety depends on the exact color-burst frequency of 3.579 545 MHz. Any rainbow generator is keyed to exactly the line horizontal rate of 15 734 Hz subtracted from the color-burst rate, which then is 3.563 811 MHz. As a result, the rainbow generator drops one cycle in each 63.5-μs completed horizontal line and blanking scan, even though it is in phase at both the start and finish of the period. Because scanning is offset by a precise multiple of the horizontal rate, the receiver can't tell the difference, and the term "sidelock" has evolved. Red, blue, and green colors are then gated by a master 188-kHz oscillator in changing phase from start to finish, producing yellow-orange, orange, reds, magenta, blues, cyan and greens—much the same (except orange and true yellow) as the NTSC standard, which produces the fully saturated colors of yellow, green, blue, cyan, magenta, and red. Nor do the sidelock generators offer red, blue, and green unicolor fields for color-temperature setup and impurity investigations. On the other hand, a good color-eye evaluation with a gated rainbow generator is often easier because of the 10 bars and the more convenient shadings.

In troubleshooting receivers, there is no parallel between the two equipment because the sidelock generator, without disruptive between-bar blanking, is much preferred; only a series of discrete voltage levels is visible via the NTSC type through an oscilloscope. The sidelock's 10 bars show as "houndstooths" of voltage at varying amplitudes and phase. Unfortunately, because all chroma-luma signals now find their way to the cathodes of the picture tubes, most vectorscopes now produce pictures 180° out-of-phase (upside down). Newer sidelock color-bar generators will, undoubtedly, correct this annoyance with an inverter. Some of the older color-bar generators no longer made, such as Sencore's CG169, will produce excellent, nondistorting color bars, but most new ones will not (as a result of interbar blanking), which shows up in the vector as tiny lines that dart through the ten waveform petals.

As for NTSC units, Tektronix, in addition to its automated transmitter signal equipment, continues to offer its lower cost. If you want to spend approximately three times as much, there's Tek's TSG170G, studio test signal generator with VITS & VIRS inserter.

As for NTSC units, Tektronix, in addition to its automated transmitter signal equipment, continues to offer its lower cost 1720 vectorscope with stereo audio display or German PAL as options. Included are X-Y inputs, channel A-B, auxiliary control and demodulator output (Fig. 10-1). Figure 10-2 is the TSG-100 NTSC test signal generator with 8-bit digital generation, 1-kHz audio tone, H/V scope triggers, sinx/x, color bars, and matrix functions. More expensive is the TSG-1700 digital composite NTSC generator with genlock, black burst, tape-leader count-down, IRE flat fields, luminance, and multibars (Fig. 10-3).

Fig. 10-2 *NTSC color and sync generator.* Tektronix

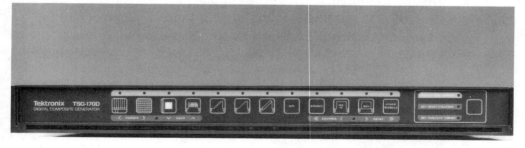

Fig. 10-3 *Tek's digital NTSC deluxe generator, including SMPTE color bars.*

Although these vectors appear on the circular graticule, polar coordinates permit phase and amplitude measurements at the various indicated points, marked in the primary and comple-mentary colors of red, blue, green, cyan, yellow, and magenta. Errors in encoding, color transmission problems, modulation or demodulation, intersystem coupling, and hue (tint) examinations of chroma phase relative to burst are all open for analysis. In very fancy NTSC vectorscopes, linear time bases also operate very precisely at line rates and color signals can be demodulated at I & Q, R-Y, U, & V (PAL), which is the German system called *Phase Alternate Line*. Luminance can also be separated from chroma and shown separately or recombined; phase matching between video cables is possible to less than 0.5°.

Although this chapter isn't about color-equipment generators per se, it's impossible to view vectors in the various video equipment without an accurate signal source, and their versatile outputs are of more than casual significance. So, we have already described some of their functions, and could well do more as the subject further develops.

How vectors originate

A vectorscope energizes quadrature color outputs following demodulation in a studio setup and between vertical and horizontal cathode ray tube exciters in an R-Y, B-Y color television receiver. Its purpose is to show you precisely what the chroma circuits are doing and possibly why—especially if you can competently interpret waveforms. In the receiver, however, B-Y (green) is not engaged because it has no quadrature kinship to the other two. The result is a two-phase relationship where demodulation between I and Q, or R-Y, B-Y (phase shifted 30° from I and Q), occurs and reconstructs the various saturated and nonsaturated colors on the face of the cathode ray tube whenever combined with Y (luminance).

We are, however, not concerned with luminance when looking at signal-generated colors. Only chroma, which is processed following the video detector and cleanly separated with either traps or comb filters from luminance. If luminance was permitted in the final display, you would find both an increase in amplitude and an irregular waveform which distorts the very information you wish to see. At the studio, of course, luminance is easily excluded. The newer television receivers have a better control of black levels, contrast, and brightness, which permit vector investigation without troublesome capacitance luma trapping that was once mandatory for recognizable displays. Most oscilloscopes worthy of the name now have X-Y circuits, which permit good lower frequency vector reproductions between 1 to 5 MHz without undue distortion. Special NTSC vectorscopes, of course, are designed for this sole purpose and do have very different internal amplifier and time-base arrangements, plus a stiff tariff to go with them.

Vectors really originate from series or parallel-tuned circuits. Unlike scalar quantities, which have mass or length, the vector has both size and direction. These quantities are characterized in terms of magnitude and phase angles. Therefore, there is no such thing as simple addition of vector quantities. All must be done algebraically, with a bit of trigonometry tossed in.

As you will recall, in series tuned circuits, impedance is minimum at resonance with maximum current flow. In parallel tuned circuits, impedance is maximum at resonance, current minimum, and the power factor equals 1. Here, however, we really don't care whether these circuits are series or parallel tuned, but we are interested in a sine wave being the vertical projection of a rotating vector. In this regard, when capacitors and inductors shape circuits with currents leading voltages and currents lagging voltages, current lead or lag becomes quite important because both phase angles and magnitudes automatically change.

If you would plot a series-tuned circuit consisting of inductance, capacitance, and resistance, the pure plot would look like that in Fig. 10-4. Notice that series inductance (reactance) is plotted in a positive direction, series capacitance in a negative direction, and resistance is perpendicular to both.

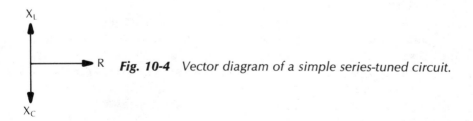

Fig. 10-4 *Vector diagram of a simple series-tuned circuit.*

Now, considering that reactances have been introduced into the circuit, the perpendicular resistance immediately becomes an impedance, with the resultant magnitude and direction totally changed. Furthermore, we want to know if the reactive element has now become more inductive or capacitative because this will phase locate the direction that the circuit is taking.

As an example, put in a few simple numbers, two reactances, and one resistance, and calculate the resultant impedance and the phase angle. In Fig. 10-5A, let's say that the inductive reactance is 5, the capacitance reactance is 3, and series resistance is 8. The impedance of this circuit is:

$$Z = \sqrt{R^2 + (X_L - X_c)^2}$$
$$\text{or} \quad Z = \sqrt{8^2 + (5 - 3)^2}$$
$$\text{then} \quad Z = 8.25$$

The phase angle is $X_L/R = $ Arc Tan, or Tan^{-1} = 2/8. The phase angle is then 14°. Therefore, the complete *polar* equation now becomes:

$$8.25 \; 14°$$

This answer turns outs to be exactly what's illustrated in Fig. 10-5B. So, you have a low-Q circuit with a small positive phase angle resulting, just as the figure would indicate. By the way, if you had been dealing with a parallel resonant circuit, calculations would have been in terms of currents, rather than straightforward impedances. This change results from current flow that's opposite in the two reactances. If a parallel circuit had no losses, voltages would always be equal and no current would flow. However, all circuits have losses and, consequently, by Ohm's law ($I = E/Z$), current becomes the dominant factor. It's also well to remember that a parallel circuit below resonance is always inductive, but above resonance it's capacitative: precisely the opposite of those in series resonance.

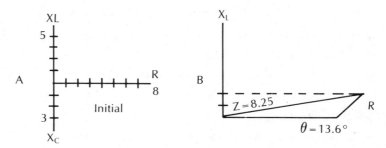

Fig. 10-5 *Differing inductive and capacitive reactances produce both positive and negative results.*

Now that the modus operandi of vectors have been established, the time is at hand to show how color television's vectors are actually formed. Do you remember *Operator J*? Please recall that:

$$j^1 = j, j^2 = -1, j^3 = -j, \text{ and } j^4 = 1$$

As each j-factor appears, it rotates the vector 90° counterclockwise, and so it passes through the four quadrants of any circle. A sine wave, of course, does exactly this, and as it rotates each of the affected quadrants anywhere between 0° and 360°, but in either partial or total 90° intervals. In monitors or color television receivers, chroma demodulated signals reach vertical and horizontal plates of the cathode ray tube at least 90° out of phase, which aid and oppose one another and place selected patterns in the four quadrants of the X and Y axes. A specific example of this appears in Fig. 10-6. Here, input signals are exactly 180° out of phase with one another, producing equivalent parts of a vector in the second and fourth quadrants. Amplitudes of each waveform, as you can see, are the same also.

In order to visualize what's happening, imagine capacitative and inductive situations, with current leading and lagging. This

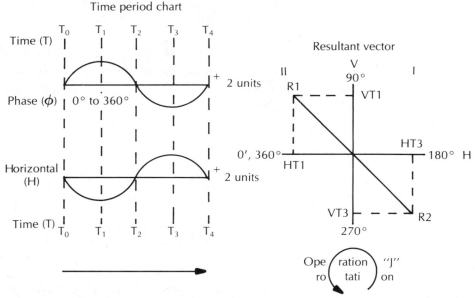

Fig. 10-6 *Equal amplitude sine waves of opposite phase produce a resultant vector in the second and fourth quadrants.*

time, they're 180° out of phase from start (T_o to finish T_2). There-fore, with time as T_1 positive maximum on the vertical axis, and HT_1 negative maximum on the horizontal axis, the first R_1 resultant appears in the second quadrant, just 45° centered between the two. Here, J^2 becomes -1.

The second R_2 *resultant begins with* T_2 and ends with T_4. As you can see in Fig. 10-6, the symmetrical sine wave goes negative for VT_3 and reaches its maximum alternation at T_3. At the same time, horizontal HT_3 goes positive, and also gains maximum amplitude at T_3. Now, because this is in the fourth quadrant, $j^4 = 1$, so VT3 and HT3 rectangle out to R2 as exactly 180° inverse of VT1 and HT1. The resultant for both vector portions becomes a straight line, but in opposite quadrants. The illustrative sine wave has come full cycle and is ready to begin another alternation. Incidentally, if you would put the same sine wave into an oscilloscope's vertical amplifiers, the straight line would appear in the first and third quadrants. This method is excellent for finding X-Y phases and amplitudes. Later, we'll demonstrate this fact.

As a further demonstration that vectors do, indeed, work and show a real story of their own, let's change both phases and amplitudes, and do an illustrative vector for Fig. 10-7. Here, the phase of each vector begins late, and the vertical portion measures only 2 units, but the horizontal portion measures 3. These

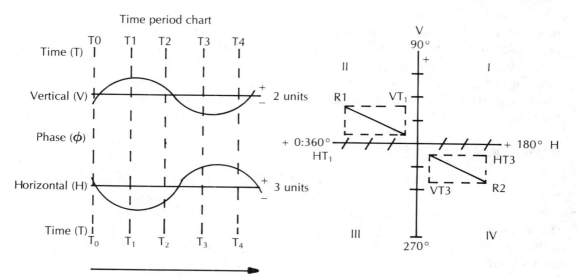

Fig. 10-7 *Out-of-phase sine waves with different amplitudes result in off-axis vectors.*

two nonsymmetrical waveforms, however, start late and phase lag so that at *T1* they are still rising in amplitude and don't attain their full value until between *T1* and *T2*. Therefore, they're offset from the X and Y axes, even though their phase is identical. Instead of 2 full units for vertical at *T1*, you find only about 1.7, and only about 2.7 for the horizontal. In the final analysis, both portions of the vector are offset from the true X-Y axes and no zero crossings are at either *T2* or *T4*, as were in Fig. 10-6.

Actually, this example is really a graphic display of complex numbers. Mathematically, these vector portions are represented by $a + jb$ and correspond to real numbers, which are points on a plane. Considering the four quadrants and their positive/negative axes (with the H axis written first), the numbers for quadrants I through IV would be:

$$+3 +2, -3 + 2j, -3 -2j, \text{ and } +3 -2j$$

Notice that only in the first and third quadrants are all signs similar, although inverted and 180° out of phase. So, $a + jb$ represents the absolute value of a complex number. To change it to a unique real number, the expression becomes:

$$a + jb = \sqrt{a^2 + b^2}$$

This polar form of complex numbers, followed by a phase angle, is represented by $a = r \cos \theta$, or $b = r \sin \theta$.

It's only in recent years that we have been able to take advantage of vectors in higher frequency applications. Before then, adequate text equipment and appropriate X-Y amplifiers were not available. Now, however, the electronic doorway has opened and you should gain maximum advantage. The easiest way to judge phase and even frequency comparisons is with a well-designed vectorscope.

Lissajous patterns

It's time now for some actual Lissajous figures that were done with two unsynchronized sine-wave generators, whose signals are put to the X-Y terminals of a dual-trace oscilloscope. Overall, these patterns look pretty good, but something is askew. Study them as the explanation progresses and see if you know the answer before we tell you.

Back in the 1800s, before oscilloscopes had been invented, Jules-Antoine Lissajous postulated the results of two sine waves in a closed line pattern. And for this discovery, such patterns have been named *Lissajous* ever since. Apparently, he was interested in these alternating voltages to determine comparative frequencies. However, we also know they are highly useful in determining phase.

In Fig. 10-8A you see two equal frequencies that produce what should be a nice, round circle. In B of the same figure, the horizontal frequency is double that of the vertical, even though it appears that there are *two vertical* alternations to one for the horizontal. In C, the vertical frequency becomes twice that of the horizontal and the two full alternations are shown in the horizontal plane of the oscilloscope. During the D portion of Fig. 10-8, the vertical frequency is tripled compared to the horizontal frequency.

True, these are good, live examples of frequency comparisons in the 1-, 2-, or 3-kHz regions. Remember that at the beginning of this particular exercise, you were told about a fault. If you haven't guessed by now, the problem is a difference in amplification between the two oscilloscope amplifiers (or input voltage, if you wish). The vertical amplifier input exceeds that of the horizontal. Consequently, the pattern is an oval, rather than a pure circle. Don't blame your oscilloscope this time, input signals are the problems, rather than any implied or suggested phase change, and they continue throughout the 4-figure display.

This "abnormality" suggests a way to check your oscilloscope doesn't it? How about a Lissajous pattern at lower and higher frequencies to check your X-Y responses? It might not be a bad idea to discover at what frequency the amplifier match falls off. Another method is to put the same sine wave into both amplifiers and increase frequencies until the resulting diagonal bar begins to shift its position or opens. At those points, the amplifiers exhibit problems and X-Y match does not continue to exist.

In Fig. 10-9A, we decided to do just that—put the same sine wave into an oscilloscope's vertical amplifiers and see what happens. At 200 kHz, as you see, the diagonal line begins to open, denoting phase shift. Not much, but enough to indicate that's about the limit of accurate phase matching for this particular piece of equipment.

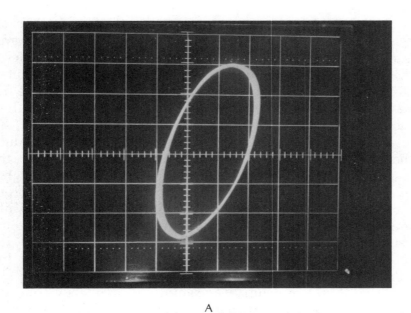

A

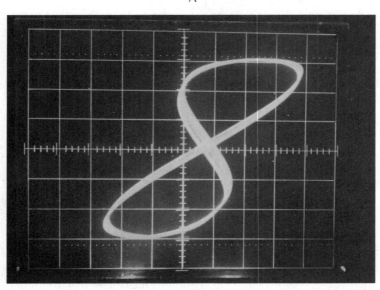

B

Fig. 10-8 *Four Lissajous patterns at same or different frequencies; but there's a problem. What is it? (A) two equal frequencies; (B) horizontal 2× vertical; (C) Vertical 2× horizontal; (D) vertical 3× horizontal.*

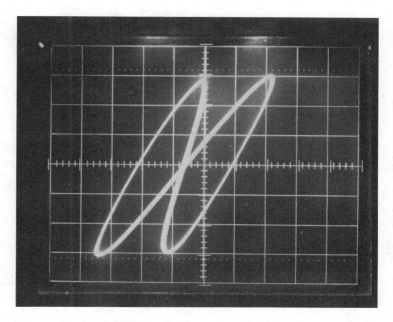

C

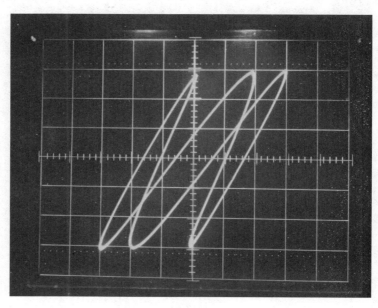

D

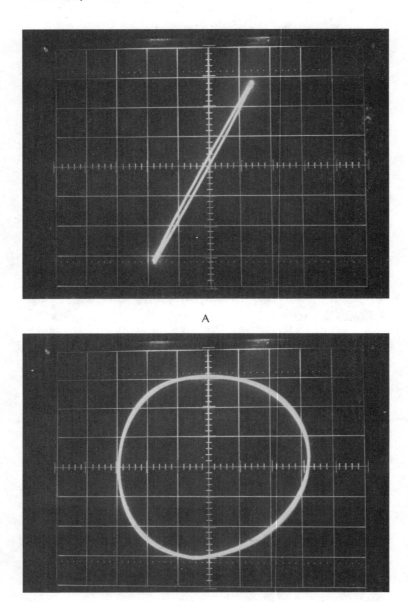

A

B

Fig. 10-9 *More Lissajous patterns testing both the scope and signal-origi-nating generator (A) amplifier phase shifts become apparent at 200 kHz. (B) an approximation of circle, with only slightly un-equal generator amplitudes (C) vertical frequency 2× horizontal (D) vertical frequency 4× horizontal.*

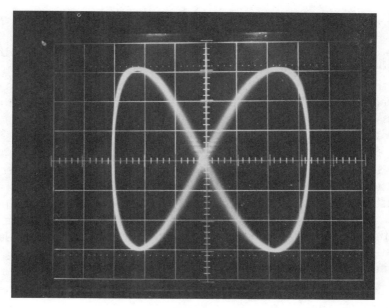

C

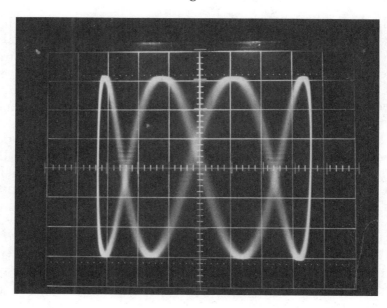

D

After that, we synchronized the two generators and produced a relatively jitter-free Lissajous circle with just a little suggestion of an oval, rather than one that's perfectly round (B). In C, the horizontal input frequency was halved so that once again the vertical frequency is double that of the horizontal. For the last of these displays (D), we reduced the horizontal input by a factor of four so that the vertical frequency is now four times that of the horizontal.

Now, all waveforms have the same amplitudes and, initially, the same frequency. They do, indeed, produce reasonable results. Naturally, other phase shifts can enter into a Lissajous pattern and twist them into all sorts of contortions. However, this short series of actual photos illustrating the uses of Lissajous patterns should suffice for now. Other phase and amplitude illustrations are later, as we look at video receivers and monitors—especially during the demodulation process.

Calculating phase shifts

At this juncture, it might be worthwhile to look at a few phase shifts to learn a useful method of calculating them. What we'll do here is to lay out several shifting signals and let them swing from right to left, crossing the various vertical and horizontal axes and from one quadrant to another. If you wish, you can use an oscilloscope graticule, mark off accurate divisions, and then make measurements as suggested. All that's involved is a little arithmetic and high school trigonometry and any calculator that will handle sine-wave trigonometric functions. You only have to draw a line from the X and Y axes (V and H) intercepts of either a diagonal or the ellipse-circles, divide one into the other and take the sine of the final result. The effort should hardly become an intellectual gyration, because the process is totally straightforward and very quick. So, let's proceed with the information in Fig. 10-10.

First, observe that only a 45° diagonal runs through (1) of the initial figure, but B equals 2. Therefore, if A is 0, then B divided into A is zero, and the arc sine of 0 is 0. In example 2, the ratio of A to B divides out to 0.5, which, when calculated by arc sine, becomes 30° and also 330°, which is 360° – 30°, because two quadrants of the four-quadrant figure are always involved.

Coming full circle and occupying all four quadrants equally, our circle in 3 divides 2/2, and the arc sine of 1 is 90° and 270°. However, the next portion becomes tricky.

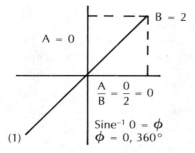

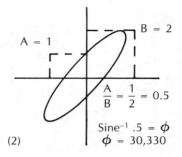

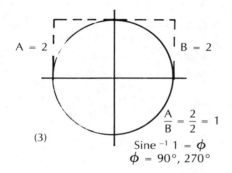

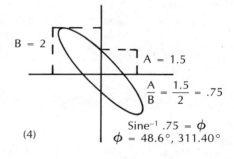

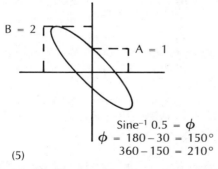

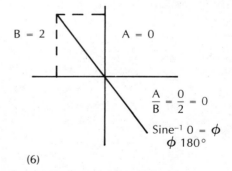

Fig. 10-10 *Phase swings from positive to negative and simple means of calculating the φ angles.*

In order to keep the arc sine value from becoming more than 1, the A, B measurement yardsticks will have to be reversed as the ellipse swings fully into the second and fourth quadrants in 4. So, A measures 1.5 and B is 2. The smaller number will have to be divided by the larger, with an answer of 0.75. The arc sine of this then becomes 48.6° and 311.4°. Still with me?

As 5, swinging, further left, becomes a mirror image of 2, you find that 0 equals 150° and 210°. And, finally, (1) and (6) also become mirror images of one another and check out similarly, except instead of 0 and 360°, (6) becomes 150° and 210°, which are simple subtractions from 180° and 360°. Isn't this an easy way to develop phase shifts? It is, of course, if you don't have microprocessor-controlled oscilloscopes to handily do such measurements for you and offer a few additional details besides. Everyone, unfortunately, can't afford $5000 or more for special esoterics.

Now that you know how to handle both frequency and phase rudiments using Lissajous circles, ellipses, and straight lines, we can now turn to prime examples of vectorscopes and what they do in actual applications. We especially want to accentuate gated rainbow generators and their results because an ordinary oscilloscope can handily display products, which are highly useful in video systems of all varieties.

Illustrating phase shifts

Waveforms taken at NBC studios in Washington, D.C. (Fig. 10-11) clearly illustrate what an NTSC vectorscope can and will do under many routine good/bad conditions. These illustrations might have been more elaborate, but both loss of amplitude and phase shifts are fairly obvious, and that's what these NTSC vectors are all about. Remember, they are basically for camera and general studio setup, *not* specifically for overall troubleshooting operations. When such are needed, an oscilloscope investigation among the various system components usually pinpoints difficulties considerably more quickly—especially when sync and luminance (Y) faults are involved. So, use these NTSC waveforms for the purpose intended (that of chroma observations) and other instruments and methods for other things (such as the video monitor, and various windows, Sine² pulses, modulated and unmodulated staircases, VITS and VIRS, etc., for characteristic problems).

As you see in (A) of Fig. 10-11, all yellow, red, magenta, blue, cyan, and green vector points are both of the required amplitude and in phase. I and Q dots are located in their respective first and fourth quadrants.

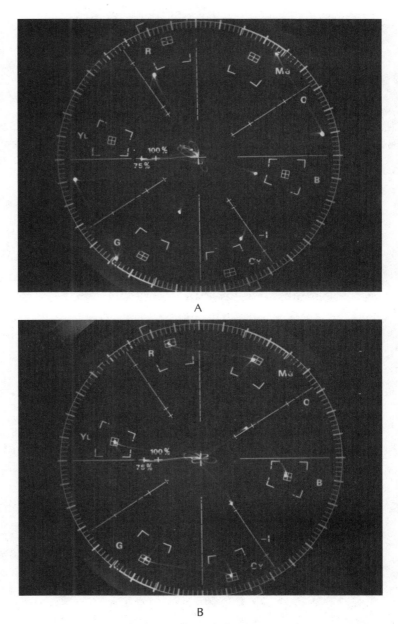

Fig. 10-11 *Good and bad NTSC vectors indicated by dot positions of six saturated colors and I/Q in the first and fourth quadrants (A) a good vector, with all six saturated colors in position (B) chroma burst phase shift (C) blue/yellow missing (D) low chroma burst amplitude.*

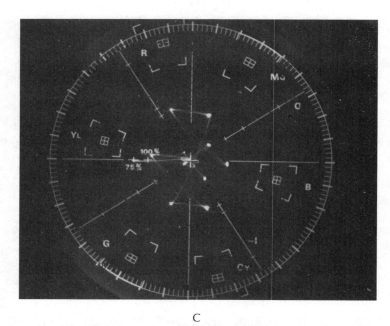

C

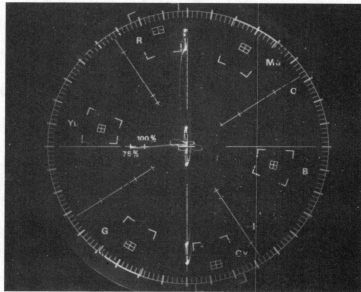

D

Fig. 10-11 *Continued*

In B, chroma burst suddenly shifts phase, people turn green or magenta at the broadcast station, and ''idiot-button'' circuits in television receivers valiantly try to spread fleshtones over twenty or thirty degrees extra demodulation angles to make up for this unwelcome difficulty. Of course, I and Q are skewed at the same time, so the receiving apparatus of varying descriptions have a hard time keeping up.

In C, the problem is radically different. If you look closely, only red and cyan remain, I and Q have disappeared, and everything is out of phase. Therefore, whatever is generating these other colors is obviously at fault and must be traced through the amplifiers of the system. Once more, receivers of all types that try to reproduce color video will react to some unusual results. You might even think you'd lost the blue gun of your cathode ray tube.

In the final (D) illustration of Fig. 10-11, both chroma and burst amplitude are down, although phase remains just about where it should. Under these conditions, you'd go directly to the burst buffer and chroma amplifiers to find the problem—ordinarily nothing more than a simple transistor or IC that can be discovered quickly by an ac/dc oscilloscope check. As those who work in broadcast maintenance will tell you, nothing is magical in troubleshooting. Just good, hard common sense and a solid rudimentary knowledge of tube and semiconductor electronics (both analog and now digital, the latter beginning to make its presence felt extensively in both audio and video applications) is necessary.

Vectors by sidelock generation

To the broadcasters, sidelock or gated rainbow vector generation isn't of overwhelming importance. They can afford highly sophisticated equipment to produce precision results and permit measurements hardly ever done in the field. But video shop people and an occasional electronic Mr. Fixit usually don't have the corporate nickels behind them, and must work with something a little less exotic than $3000 to $8000 equipment. Even today, a good sidelock color bar generator is priced around $200 and can be highly useful in all sorts of ways—especially if the operator bothered to become totally familiar with its uses.

Having been the route of so many who have gone before and who will, undoubtedly, pass this way again, a working explanation of what's involved should be of help to future generations of those struggling anew with video cassettes, monitors, and just plain old color televisions. With a clean, reliable sidelock generator, you can do anything in chroma circuits in the way of investigation and repair. In addition, the best-looking color-bar displays that you can possibly produce are the 10 bars that can be projected on the face of any color display device that has the necessary ingredients of a TV receiver. Be it baseband or RF, good colors in their various hues are going to show if the generator and the receiver/monitor are in good working order. If not, the 10 or 11 gated color bars can find the trouble every time.

The reason that NTSC-generated patterns are not especially useful in consumer-type electronics are the variable customer controls on the front panels. Unfortunately, as chroma, hue, brightness, and sometimes contrast controls are twiddled, NTSC patterns can be changed radically, and most useful data lost (Fig. 10-12). In A are the R-Y and B-Y levels that you see at the cathode ray terminals of receiver equipment with the NTSC generation. In B is the vector pattern, which shows the six "saturated" colors plus black and white. Now, move tint, contrast, color, or brightness controls and see what happens. You'll find it exceedingly difficult to keep the various colors straight, in addition to their several, varying amplitudes, if the front-panel controls are moved for any reason. This is precisely why we do not generally recommend NTSC generators to supply such drive signals for consumer electronics equipment. You can be over your head in a sea of problems before you know it and, in the end, either regular sidelock color bars or a gated rainbow vectorscope will have to pull you out. With those admonitions behind us, let's go on with the business at hand.

The gated or sidelock color bar generator was originally just a red, blue, and green display device showing only the three basic TV colors (as opposed to those regarded by printers and painters as magenta, cyan, and lemon yellow), which are formed by the subtractive rather than the artists' additive process. This crystal-controlled signal oscillation operates at 3 563 812 Hz, and is gated by a 188.8-kHz oscillator, so that during each horizontal

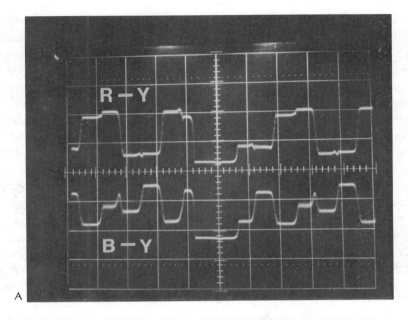

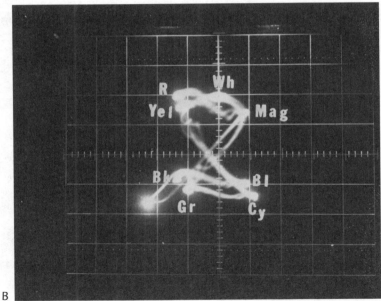

Fig. 10-12 *NTSC voltage levels and resultant vector at a TV receiver's cath-ode ray tube (A) R-Y, B-Y voltage levels; (B) NTSC vector on ordinary graticule.*

scan line, 12 bursts of color frequency energy is developed and projected through the generator's video output or RF modulator. The 12th burst, however, takes place during horizontal sync periods and is lost (Fig. 10-13). The color receiver also inserts its own 11- to 12-ms blanking interval following 52.4 ms of horizontal scan, and so another color burst is lost in the chroma circuits themselves. Therefore, when we finally look for a gated rainbow vector at the cathode ray tube, only 10 bursts of voltage, or vector petals, can be seen. However, the information gained from this 10-petal gated rainbow vector is astonishing.

Figure 10-13 pretty well tells the entire story in a composite diagram, which shows a single line of horizontal scan followed by successive illustrations of what the color-bar generator and receiver are doing simultaneously. Total scan and blanking intervals remain for 63.5 μs. During that time, all color bursts are formed and processed by the receiver, just as you see diagrammed. As each burst series interrupts the overall display, the

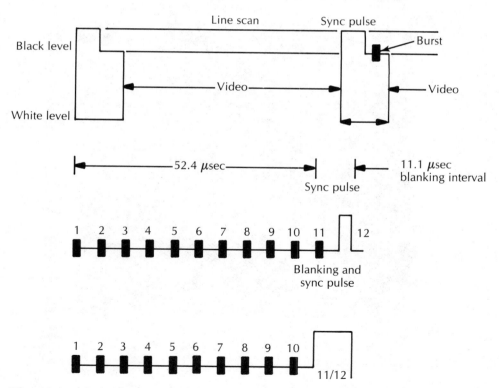

Fig. 10-13 *Color bar generator develops 12 color bursts during each horizontal scan, but 2 are lost during horizontal blanking.*

receiver's 3.58-MHz oscillator locks up through its automatic phase control system on the sync burst, which is then followed at 30° intervals by the remaining bursts until all 10 colors are visible on the receiver's cathode ray tube. This is shown in Fig. 10-14, where colors and burst phase are keyed and numbered 1 through 10. Once again, receiver blanking between 0° and 300° is illustrated in both this figure and Fig. 10-13, so you can mentally correlate the two. What's more, the phase relationship

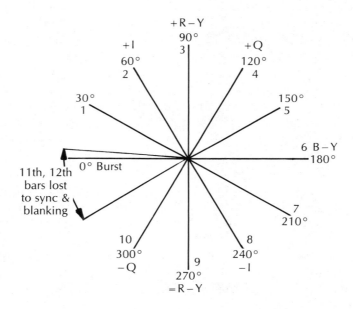

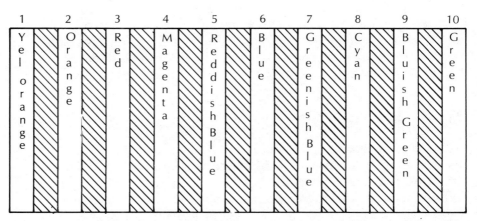

Fig. 10-14 *Gated rainbow vector shows 30° separation between petals as the final 10 color bars are developed.*

between R-Y and B-Y is plainly shown to be at quadrature (90°). The actual vector phase angle will depend on both phosphors and phase loss or gain through final chroma amplifiers and the demodulator.

You might also wonder how green is displayed in both NTSC and gated rainbow vectors, because only I and Q or R-Y and B-Y are connected to the vectorscopes. It's not really a very long story. The 1.5-MHz I signal carries color hues that range from bluish-green (cyan) to orange, while 0.5-MHz Q intelligence passes colors between yellow-green and purple. Observe, please, there is green in both I and Q, and R-Y and B-Y are just phase diffrences of some 30° over I and Q. The green gun in video receivers actually receives a mix (matrix) of R-Y and B-Y to excite its green cathode. It thereby supplies a true RGB nonsignal input raster that, when the guns are firing proportionally, produces a greyish white or white raster that will measure between 6 500° and 8 000° Kelvin, depending on the manufacturer. This raster is used for convergence checks, while individual red, blue, and green color fields are selected for color purity.

We could show you another color timing chart that might make the rainbow gating process a little clear. However, if you will simply remember that only 10 color bars are involved, the R-Y portion has a null or zero crossing at the 6th bar, and that B-Y has zero crossings at the 3rd and 9th bars, the resultant vector is virtually automatic.

Now, with all this combination of theory and electronic history stored in ample memory, let's stash some more in another register that you can use as a convenient ROM any time you choose for both reference and precision results. Many things can be done with accurate color bars to provide a great deal of highly useful information. We now, therefore, are going to begin practical applications and continue this for the remainder of the chapter.

Applications

Just to prove that sidelock generators will indeed work in video cassette recorders, Fig. 10-15 illustrates both the 11 bars of color burst (top trace) as well as the voltage in the bottom trace, being modulated with rectangular pulses, possibly from another portion of the chroma system because these pulses are obviously appearing in time with the 11 burst above. Here, a little extra

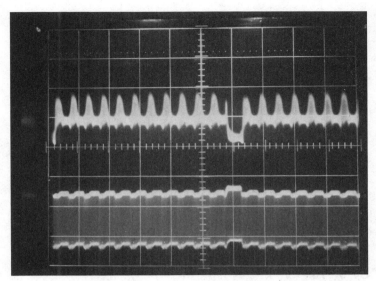

Fig. 10-15   *Gated rainbow generator signals in a VCR.*

blanking between color bars wouldn't hurt because they cannot be eventually used in vector applications anyway. If the service engineer or technician knows his equipment, either NTSC or sidelock will do, although many waveforms throughout the VCRs are shown by NTSC signals alone—at least on the schematics.

In the television receiver, however, the sidelock generator is clearly superior in all applications after the video detector, and can do a remarkable job telling all sorts of chroma tales. The 11, then 10 color bursts do their jobs throughout all chroma circuits, and then are literally chopped into 10 final color bars by the action of synchronous demodulators (Fig. 10-16) gate at the chroma subcarrier oscillator rate of 5.579 545 MHz. As two phase-shifted sine waves gate the chroma information, petals of chroma in varying amplitudes in the R-Y and B-Y demodulations are formed, with R-Y nulling (going to zero) at the 6th bar, and B-Y passing through zero at both the 3rd and 9th bars. For proper color demodulation to be affected, these color-bar amplitudes and nulls must be continuously maintained (at least while the generator is operating) so that color will be uniformly placed in the large areas of any picture where it belongs.

The sort of vector you should be looking for appears in Fig. 10-16. No crossovers are in the pattern, and it's reasonably symmetrical, although petals 5 through 10 are a little fuller than we

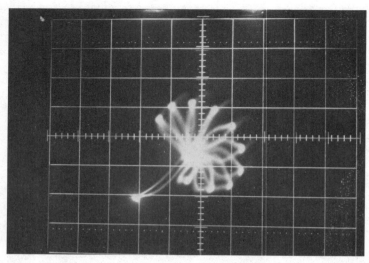

Fig. 10-16   *For today's solid-state receivers, this is good vector.*

would like. The entire display is highly serviceable, nonetheless, and it is resolved directly from the "tooths" of demodulation (Fig. 10-17).

If this pattern whirls in a tight circle, your 3.58-MHz chroma oscillator is out of sync. If the bottom and top amplitude collapse, problems are in the red amplifier. If the pattern reduces in amplitude symmetrically, chroma bandpass amplifiers are deficient. If a side collapses, the blue final amplifier (after demodulation) shows fault. If the pattern becomes twisted and crossovers appear in the petals, either your bandpass amplifiers are not properly compensated or aligned, or the problem is with fine tuning the VHF or UHF tuners.

The third bar from the left (the R-Y bar) should have relatively fast risetimes and falltimes, and the usual angle of demodulation between the third and B-Y sixth bars normally is 90°. If an "idiot button" is depressed, a closing of the first three bars is common practice and the angle of demodulation often increases from 105° to 120°. This, of course, is not so in the VIR and Color-Pilot II receivers, because they use line 19 of the blanking interval to pick up VIRS and then correlate red and blue color levels between the receiver and transmitter for precise chroma rendition.

With the relaxation of broadcast standards by the FCC, including removal of the First Class Radiotelephone require-

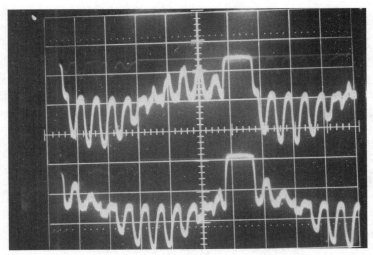

Fig. 10-17 *"Hounds tooths" of chroma demodulation place color bars in accurate phase over picture.*

ment, more VIR receivers are needed in all lines to control broadcast color variations of all descriptions. Some of the older VHF stations, however, are now installing both new and more automated equipment for colorkeeping, as well as circular polarized antennas that should help reduce "ghosting," so some of these problems might, in time, become somewhat less annoying.

Summary

In summarizing the various material and thoughts that have been developed during this chapter, remember that a *scalar* quantity can have volume, mass, density, etc., and it can be arithmetically added. A *vector*, on the other hand, is a complex quantity described in rectangular coordinates as $a + jb$, it has both distance and direction, and it revolves within the four quadrants of any circle, according to the counterclockwise motion of operator "*j*."

Vectors are useful in both relatively low frequency comparative measurements as well as in phase angles. NTSC types are especially convenient for broadcast cameras and studio setups. Gated rainbow generators and their resultants work much better with most consumer video products, especially television receivers. Remember, however, there must be either an I and Q or R-Y, B-Y set of connectors for these vectorscopes to monitor if you want to see a vector. Videocassette recorders just upbeat and downbeat FM chroma without demodulating, so there are no I/Q, RB-Y connections.

You are also again cautioned about choosing a useful rainbow color bar generator. Many, or most, have special blanking circuits that do, unfortunately, destroy the effectiveness of the 10-petal vector display, place too much interference in the vector, and make it virtually unusable. Certainly, the internal vector portion won't be worthwhile, and only the petal outlines can be effective. Something else is worthwhile knowing, too: neither the average service oscilloscope nor the vectorscope can invert either outputs or inputs. So, the vector is upside down, which places an extra burden on the operator, who either has to stand on his head, turn the vectorscope around, or go into other unsatisfactory contortions. All this results from prior design of older vectorscopes, which do not take into account the 180° phase reversal between cathodes of the newer picture tubes and control grids of the older varieties. We would hope some of the newer sidelock generators have polarity reversal features to accommodate all types of receivers. It would certainly help!

Finally, we do sincerely hope that applications and theory in this chapter can be followed and the lagging use of vectors and vectorscopes will be revived especially in consumer products. With complexity of ICs and equipment functions ever increasing as new products are more frequently introduced, both swept chroma and vectors are needed to analyze and troubleshoot all sorts of color circuits. The more proficient one becomes in their use, the easier both design and trouble problems can be solved. You owe it to yourself to become a student of these chroma uses. They save a great deal of time and money!

Index

A

accuracy factors, 8
accuracy/precision, 1
active filters, 40-41,
ADAC converters, 15
Advantest R4131D spectrum
  analyzers, 131-132
aliasing, digital-storage
  oscilloscopes, 10, 100
amplitude measurement,
  coupling circuits, 39
Anritsu MS610B/J spectrum
  analyzers, 132-133
apparent power, sine wave, 29
ASCII codes, 250-251
asynchronous data transfer, 255
asynchronous noise, 262
auto-test devices, 1
autodialers, 233
average power, sine wave, 28
averaging, digital-storage
  oscilloscopes, 100

B

backscatter optical
  reflectometers, 221
backscatter, fiberoptics, 223
bandpass filters, 42-43
bandwidth requirements, 7-8
baseband analysis, spectrum
  analyzers, 145
binary numbers, signature
  analyzers, 247-248
binary-coded decimal (BCD)
  numbers, signature

analyzers, 248
bit stream errors, 4, 5, 120
  logic analyzers, 178
BK-Precision PL 5610 spectrum
  analyzers, 130-131
blanking signals, 45-47
Boole, George, 173
Boolean algebra, 173
buses, communications,
  233-234
  FutureBus, 234
  FutureBus + , 243-244
  Multibus II, 234
  NuBus, 234

C

cable radar (see time-domain
  reflectometry)
calibration/recalibration, 8, 79
cameras (see also
  printers/plotters), 23-24
carrier-to-noise ratio
  correction factors, AM and
    FM, 139
  signal-to-noise ratio, 138-140
cathode-ray tubes, 7
clock speed, logic analyzers,
  174-175
color burst, NTSC signal, 44
communications standards,
  232-234
  autodialers, 233
  buses, 233-234
  FutureBus, 234
  FutureBus + , 243-244

GPIB standards, 232-234
IEEE-448 standards, 232-234
Multibus II, 234
NuBus, 234
protocols, 233
termination characters, 233
VMEbus, 234
VXIbus, 233-234
compaction, digital-storage
  oscilloscopes, 102
composite signatures, 263
compression point, spectrum
  analyzers, 159
computer interfaces, 230-231,
  235-236
conversions, number systems,
  264
counter-timers, 1
coupling circuits, 37-40
  amplitude, 39
  conductive configurations,38
  efficiency, 40
  filters vs., 41
  impedance, 39
  LC circuits, 38-39
  maximum transformer
    efficiency, 40
  nonconductive
    configurations, 38
  phase shift, 39
  power loss, 39
  RC circuits, 38-39
  transformers and impedance,
    39-40
current, sine wave, 28
cursors, 1

D

data point spacing, optical
reflectometers, 226
data transfer processes, 254-255
data-acquisition systems,
234-235
dead zones, fiberoptics, 226
debugging, 2
logic analyzers, debugging
hard- and software, 13, 174
decimal radix numbers,
signature analyzers, 246-247
demodulation, modems,
255-256
deMorgan's Theorem, 173
design, 3-4
differentiation, 33-35
digital circuit testers, 256-262
digital readout, 5
digital-storage oscilloscopes
(DSO), 8-10, 96-125
accuracy/precision of
measurement, 100-101
aliasing, 10, 100
amplitude measurement,
117-118
analog scopes vs., 10
analog-to-digital converters
(ADC), 99
applications and uses,
103-104
averaging, 100
bit stream errors, 120
block diagrams, 121-123
cathode ray tubes, 9
compaction, 102
cost and performance curves,
10, 123-124
developmental history,
97-100
display systems, 100, 121
duty cycles, 118
equivalent time sampling, 98
event capture-and-hold
capabilities, 9, 99
Fast Fourier Transforms
(FFT), 103
frequency measurement, 118
glitch capture, 101-102
glitch triggers, 102
glitches, 115-116, 119-120
Hewlett-Packard

54600A/5460 (see
Hewlett-Packard
54600A/54601A)
horizontal loss, 102
linear interpolation, 104
manufacturers of DSOs, 97
memory expansion, 98
nonlinear interpolation, 104
Nyquist sampling rate, 10, 99
one-shot capture, 106,
115-116, 119-120
operation of DSO, 9
passband requirements, 104
periods, 118
pretrigger storage, 98
price range, 97-98
printer/plotter coupled to
DSO, 96-97
programmable, 121
quantizing errors, 103
resolution, 100-104
roll-mode screen display
updating, 98
sampling procedures, 97
sampling rates, 9, 97-99, 121
spectrum analyzers vs.,
14-15
stored writing speeds, 98
Tektronix 2232 (see Tektronix
2232 DSO)
time bases, 121
troubleshooting, 124-125
useful bandwidth (USB), 10,
99
vertical resolution, 100
display dynamic range,
spectrum analyzers, 158-159
display systems, 1
cathode-ray tubes, 7
digital-storage oscilloscopes
(DSO), 100, 121
future developments, 229
graticule lighting, 7
no picture, 4
oscillocopes, 80
spectrum analyzers, 127-128
distance-to-discontinuity,
time-domain reflectometry,
195
distortion
modulation, 160
spectrum analyzers, 18
Dolby Delta Surround

multichannel sound (MTS)
system, 157
double negation, 173
drive voltages, 4
duty cycle
digital-storage oscilloscopes
(DSO), measurement, 118
pulse circuits, 37
square wave, 30-31, 64

E

EBCDIC codes, 250, 252-253
effective power (see average
power)
efficiency of transformers, 40
encryption, fiberoptics, 218
Engleson, Morris, 158-159
Epson FX-850 printer/plotter,
24-25, 75-79, 92-95
equivalent time sampling,
digital-storage oscilloscopes,
98

F

falltime measurement,
oscilloscope, 7, 87-88
Fast Fourier Transform (FFT)
analyzers, 16
Fast Fourier Transforms (FFT),
digital-storage oscilloscopes,
103
fault-finder optical
reflectometers, 221-222
fiberoptics (see also optical
reflectometers), 218-221
backscatter ODTRs, 221
bundled cable, 218
cladding, 218
connections and connectors,
220-221
core, 218
dead zones, 226
detectors, photodetectors, 219
encryption, 218
fault distance calculation,
222
fault-finder ODTRs, 221-222
Fresnel reflections, 223
installation techniques, 221,
226-228
intermittents, 227
loss producing components,
222

multimode cable, 218
open tips, 228
Rayleigh backscattering, 223
readout resolutions, ODTRs, 226
rings, 228
single-mode fibers, 218-219
speed of transmission, 219-220
telephone faults, 227
terminal fault locators, 228
tracer fault locators, 228
transmission towers, 227
wirebond connections, 227
filters, 40-43
active filters, 40-41
bandpass filters, 42-43
bandstop filters, 42-43
coupling circuits vs., 41
highpass filters, 41-43
lowpass filters, 41-43
passive filters, 40-41
surface-wave acoustical filters
(SWIFS), 42
Fluke/Philips PM3580-family
logic analyzers, 181-193
channel
connections/availability, 187
clocking, 190-191
data labels, 187
disassembly, 193
dual sequencers, 187
help screens, 182
logical separates, 187
operation, 182-183
performance and options,
181-182
scope or analyzer use, 188
self-test and de-skew settings,
187
setup and hold time, 186
state memory, 186
state or timing modes, 188-190
state speeds, 182, 186
timing and state readouts,
183-186
timing resolution, 187
trigger levels, 187
triggering, 187, 190-191
frequency measurement
digital-storage oscilloscopes,
118
oscilloscope, 87-88
frequency response
AM radio signals, 151

FM stereo signals, 151-154
multichannel sound (MTS)
receivers, 151, 154-158
spectrum analyzers, 150-154
Fresnel reflections, fiberoptics,
223
fundamentals (see sine waves)
future developments, 229-244
autodialers, 233
buses, communications
buses, 233-234
communications, standards,
232-234
data-acquisition systems,
234-235
display systems, 229
FutureBus, 234
FutureBus +, 243-244
GPIB interface and
communication standards,
232-234
graphic user interfaces (GUI),
230-231
hi-resolution I/O board for
Mac II (NB-A2100), 236-239
IBM interface, PS/2, 235-236
IEEE-448 communications
standards, 232-234
interfaces, computer
interfaces, 230-231, 235-236
LabVIEW, 239-241
LabWindows, 241-243
Multibus II, 234
NuBus, 234
protocols, 233
robotics, 229-230
SCPI syntax/structure
language, 233
software developments, 231,
239-243
standards, communications
standards, 232-234
storage devices/media,
231-232
VMEbus, 234
VXIbus, 233-234
FutureBus, 234
FutureBus +, 243-244

G

gated rainbow vectors (see also
vectors/vectorscopes), 84-86
glitch capture, DSOs, 101-102

glitch trigger, DSOs, 102
glitches, DSOs, 115-116,
119-120
GPIB interface and
communication standard,
232-234
graphic user interface (GUI),
230-231
graticule lighting, 7
ground loops, 262
grounding oscilloscope, 79-80
GSTAR 4 satellite, 142-145

H

Hameg Corp. 500-MHz
spectrum analyzers, 129-130
Hameg Corp. HM 604 time
base oscilloscope, 66-70,
81-88
Hameg Corp. oscilloscopes, 57
hardware debugging with logic
analyzers, 13
harmonic distortion, spectrum
analyzers, 18
harmonics
sine wave, 28
square wave, 31
HDTV technology, 21-23, 26
heat, 5
Hewlett-Packard 1652B logic
analyzers, 176-180
add-on options, 177
bit stream errors, 178
delay times, 178
minimum-to-maximum
waveform values, 178
one-shot capture, 178
printouts, 178
pulse measurement, 178
sampling rates, 179
triggering in state analysis,
179-180
voltage measurement, 178
Hewlett-Packard
54600A/54601A DSO,
113-124
accuracy/precision, 113
amplitude measurements,
117
bit stream errors, 120
channels, 113
cursor setting, 115
display modes, 114

Hewlett-Packard 54600A/54601A
DSO (*cont.*)
  display system, 115
  glitches, 115-120
  HP Think Jet printer, 116
  one-shots/one-shot capture,
    115-116, 119-120
  parameter setting, 117
  period, duty cycles, 118
  probes, 114
  resolution, 113
  rotary pulse generators, 115
  sampling rates, 113
  size and portability, 114
  special measurements,
    special settings, 115
  storage mode, 117
  triggers, 113, 115
  voltage measurement, 115
Hewlett-Packard HP 3563A
  spectrum analyzers, 20-21,137
Hewlett-Packard HP 5006A
  signature analyzers, 256-261
  accuracy, 260-261
  noise, 262-263
  signature characters and
    operation, 257
Hewlett-Packard 7090A
  printer/plotter, 24
Hewlett-Packard HP 8590B
  spectrum analyzers, 135-136
Hewlett-Packard Model 83620A
  spectrum analyzers, 168-170
Hewlett-Packard Think Jet
  printer, 116
hexadecimal numbers, 249-250
hi-resolution I/O board for Mac
  II (NB-A2100), 236-239
high-frequency equipment, 5
highpass filters, 41-43
horizontal loss, digital-storage
  oscilloscopes, 102
hum in pulse circuits, 36
hyperbolic functions, 32-33

I

IBM PS/2 interface, 235-236
IEEE-448 communications
  standards, 232-234
IFR Model A-7550 spectrum
  analyzers, 132-133
impedance
  characteristic impedance,

transmission lines, 200
  complex impedance,
    transmission lines, 199
  coupling circuits, 39
  input impedance,
    transmission lines, 196-198
  lossless (ideal) transmission
    lines, 199
  sine wave, 30
  transformers, 39-40
index of modulation, 160-161
input, 4
instrument capabilities, 2, 6-8
integration, 33-35
interfaces, computer interfaces,
  230-231, 235-236
intermittents, fiberoptics, 227
intermodulation distortion,
  spectrum analyzers, 159
IRE references, NTSC signal, 45
ITT Intermetal Semiconductor
  Digit 2000 program, MTS,
  155

J

j operators,
  vectors/vectorscopes,
  272-273
jitter in sawtooth wave, 92-93
John Fluke Manufacturing,
  logic analyzers, 176

K

Karnaugh maps, 173

L

LabVIEW 2 software, 231, 234,
  239-241
LabWindows software, 231,
  234, 241-243
launch ports, optical
  reflectometers, 222
LC circuits, coupling circuits,
  38-39
linear tracing, 6-7
liquid-crystal displays, 1
Lissajous patterns,
  vectors/vectorscopes,
  274-280
loads, 5
logic, contamination of logic, 4
logic analyzers, 11-13, 172-193

add-on options, attachments,
  176, 177
applications and uses, 11,
  172-175
bit stream errors, 178
Boolean algebra concepts,
  173
channels, Fluke/Philips
  PM3580, 187
clock speed, 12
clocking, Fluke/Philips
  PM3580, 190-191
code translation capability,
  174
cost vs. performance options,
  175
data acquisition/analysis,
  Fluke/Philips PM3580,
  191-193
data labels, Fluke/Philips
  PM3580, 187
de-skew, Fluke/Philips
  PM3580, 187
debugging
  hardware/software, 13, 174
delay times, 178
disassembly, Fluke/Philips
  PM3580, 193
disassembly-display mode,
  12, 193
display systems, 174
dual sequencers,
  Fluke/Philips PM3580, 187
Fluke/Philips PM3580-family
  logic analyzers, 181-193
Hewlett-Packard 1652B logic
  analyzers, 176-180
John Fluke Manufacturing,
  176
logical separates,
  Fluke/Philips PM3580, 187
memory capacity, 12, 175
minimum-to-maximum
  waveform values, 178
one-shot capture, 178
operation, 11
post-trigger mode, 12
pre-trigger mode, 12
printouts, 178
programmable logic devices,
  172-173
pulse measurement, 178
recent technological

advances in logic analyzers,
  175-176, 181
sampling rates, 179
self-test, Fluke/Philips
  PM3580, 187
setup and hold times,
  Fluke/Philips PM3580, 186
setup and use, 176
skew specifications, 12
state functions, 12
state memory, Fluke/Philips
  PM3580, 186
state mode analysis,
  Fluke/Philips PM3580,
  188-190
state speeds, Fluke/Philips
  PM3580, 186
time measurement/clock
  speeds, 12, 174-175
timing mode analysis,
  Fluke/Philips PM3580,
  188-190
timing resolution,
  Fluke/Philips PM3580, 187
triggering in state analysis,
  179-180
triggering, Fluke/Philips
  PM3580, 187
triggering, Fluke/Philips
  PM3580, 190-191
triggers, 11-12
vectors, 13
voltage measurement, 178
logical numbers, signature
  analyzers, 245-246
lowpass filters, 41-43
LR circuits
  instantaneous current, 32
  instantaneous voltage, 32
  time constants, 32

M

Mac II, hi-resolution I/O board
  (NB-A2100), 236-239
marker resolution, optical
  reflectometers, 226
max scan mode, spectrum
  analyzers, 19
maximum transformer
  efficiency, 40
MC-GPIB interface, 235-236
memory expansion,
  digital-storage oscilloscopes, 98

millirho measurement,
  time-domain reflectometry,
  215-217
modems, 255-256
modulation, 159-162
  AM modulation percentage,
  159-160
  distortions, 160
  modems, 255-256
  modulation index
  calculations, 160-161
  total harmonic distortion
  (THD), 160
modulation index, 160-161
Multibus II, 234
multichannel sound (MTS)
  Dolby Delta Surround, 157
  frequency response, 151,
  154-158
  ITT Intermetal
  Semiconductor Digit 200
  program, 155
  spatial equalization (SEq),
  157
  Zenith ADC 2300 U, 155-157
  Zenith APU 2400 U, 155-157
  Zenith BTSC-dBx system,
  154-158

N

NB-A2100 hi-resolution I/O
  board for Mac II, 236-239
needs analysis, instrument
  capabilities, 2, 6-8
noise
  asynchronous noise, 262
  carrier-to-noise ratio, noise
  component, 145
  noise figure (NF), 145
  signal-to-noise ratio, noise
  component, 138, 145
  signature analyzers, 262-263
  spectrum analyzers, 18, 146
  spectrum analyzers
  measurements, 145-150, 145
  synchronous noise, 262
  total harmonic distortion
  (THD), 146-148
noise figure (NF), 145
nonloading instrumentation, 5
NTSC signals (see also
  vectors/vectorscopes), 44-47
  color burst, 44

IRE references, 45
vectors/vectorscopes, 265-269
NuBus, 234
Nyquist sampling rate, DSOs,
  10, 99

O

octal numbers, signature
  analyzers, 248-249
on-screen readouts, 1
one-shots, DSO capture,
  115-116, 119-120
optical reflectometers (see also
  fiberoptics), 217-228
  applications and uses,
  223-225
  backscatter ODTRs, 221
  basic procedures and
  operations, 217
  data point spacing, 226
  dead zones, 226
  fault distance calculation,
  222
  fault finders, 221-222
  fiberoptic specifics, 218
  Fresnel reflections, 223
  intermittents, 227
  launch ports, 222
  loss producing components
  in fiberoptics, 222
  marker resolution, 226
  open tips, 228
  Rayleigh backscattering, 223
  readout resolution, 226
  rings, 228
  techniques and tips for use,
  226-228
  telephone faults, 227
  terminal fault locators, 228
  tracer fault locators, 228
  wirebond connections, 227
oscilloscopes, 54-95
  ac/dc combinations, 85
  amplitude measurements, 62
  analog scope with digital
  readout, 88-95
  applications and uses for
  oscilloscope, 79-95
  audio measurements, 85
  average value of sine waves,
  63
  basic components and
  operation, 57-65

oscilloscopes (*cont.*)
  calibration, 79
  cathode ray display systems, 80
  channel switching, 61
  circuit loading, 79-80
  developmental history, 54-55
  display modes, 59, 61
  display systems, 80
  divisions/crosshatches on display, 80
  effective voltage of sine waves, 62
  Epson FX-850 printer/plotter (*see* Epson FX-850 printer/plotter)
  falltimes, 87-88
  frequency measurement, 87-88
  grounding of oscilloscope, 79-80
  Hameg Corp. HM 604 time base oscilloscope, example, 81-88
  Hameg oscilloscopes, 57
  instantaneous values of sine waves, 62
  measurements possible using oscilloscope, 80
  off-scale warnings, 87
  operation of basic oscilloscope, 55-57
  phase-difference measurements, 61-62, 85
  photography, camera attachment, 83-84, 83
  power supplies, power supply regulation, 80-81
  printer/plotter-type oscilloscope, 70, 75-79, 92-95
  probe adjustments, 56-57, 59, 61
  probes, 55-57, 79
  programmable-function oscilloscope, 70-75, 88-95
  pulse analysis, 55, 63
  rainbow generator use, gated rainbow vectors, 84-86
  recent technological advances, 54-55
  rectangular wave analysis, 63-64
  RF signals, incoming, 81-82
  risetimes, 87-88
  rms values, 55
  sine-wave analysis, 85
  spurious pickups, 79
  square-wave analysis, 55, 63-65, 85
  Tektronix 2252 (*see* Tektronix 2252 programmable oscilloscope)
  test patterns, test pattern abnormalities, 83
  time bases, 65-70
  time inversions, 85
  vector analysis, 84-86
  vertical blanking interval (VBI), 81-83
  vertical interval reference signal (VIRS), 81-83
  vertical interval test signal (VITS), 81-83
  X/Y board, 59-60
  Y controls, 58-59
output, 4
overload, 5
overranging, 5
overshoot in pulse circuits, 36

P

padders, time-domain reflectometry, 204-205
parallel-filter analyzers, 16
passband requirements, 6-8
passive filters, 40-41
periods, digital-storage oscilloscopes, 118
phase, sine wave, 28
phase alternate line (PAL) systems, vectors/vectorscopes, 269
phase shift, 265
  calculation, vectors/vectorscopes, 280-282
  coupling circuits, 39
  illustration, vectors/vectorscopes, 282-285
  oscilloscope measurement, 61-62, 85
plotters (*see* printers/plotters)
polar equations, vectors/vectorscopes, 271

Polaroid DS-34 oscilloscope camera, 24, 96
portability, 1
power
  pulse circuits, 37
  sine wave, 28, 29-30
  square waves, 64
power factor, sine wave, 29
power loss, coupling circuits, 39
power supplies, 4
  oscilloscope, 80-81
preshoot in pulse circuits, 36
pretrigger storage, digital-storage oscilloscopes, 98
price vs. performance, 2
printers/plotters, 23-26
  DSO coupled to printer/plotter, 96-97
  Epson FX-850 printer/plotter, 24-25
  Hewlett-Packard 7090A printer/plotter, 24
  HP Think Jet, 116
  oscilloscope with built-in printer/plotter, 70, 75-79, 92-95
  Polaroid DS-34 oscilloscope camera, 24, 96
probes, oscilloscope, 55-57, 79
programmable-function oscilloscope, 1, 70-75, 88-95, 121
protocols, 233
PS/2, IBM interface, 235-236
pulse circuits, 36-37
  duty cycle, 37
  hum, 36
  overshoot, 36
  power, 37
  preshoot, 36
  ringing, 36
pulse measurement
  logic analyzers, 178
  oscilloscope analysis, 55, 63
  time-domain reflectometry measurement, 194

Q

quantizing errors, digital-storage oscilloscopes, 103

R

rainbow generators (*see also*
    vectors/vectorscopes), 84-86
Rayleigh backscattering,
    fiberoptics, 223
RC circuits
    coupling circuits, 38-39
    differentiation, 33-35
    instantaneous current, 32
    instantaneous voltage, 32
    integration, 33-35
    time constants, 32
reactance
    transmission lines, large
        reactances, 201-204
    transmission lines, small
        reactances, 204
readout resolution, optical
    reflectometers, 226
rectangle wave
    average value, 64-65
    frequency measurement, 65
    oscilloscope analysis, 63-64
recurrent sweep oscilloscopes, 1
reflectometry (*see* time-domain
    reflectometry)
rental units, 2
resistive discontinuities, 201
resolution, 1, 100-104
ringing in pulse circuits, 36
risetimes, 1, 7, 87-88
rms power (*see* average power)
rms values, oscilloscope
    analysis, 55
robotics, 229-230
roll-mode screen display
    updating, digital-storage
    oscilloscopes, 98
rolloff factors, 8

S

sampling rates
    DSOs, 9
    logic analyzers, 179
Santori, Michael, 242
satellite transmission
    carrier-to-noise ratio, 138-145
    dBm power signal level,
        140-142
    DBW power levels, 142
    GSTAR 4, sample problems,
        142-145

signal-to-noise ratio, 138-145
sawtooth wave, jitter, 92-93
scaling, automatic, 6
scanning per division mode,
    spectrum analyzers, 19
SCPI syntax/structure language,
    233
series-tuned circuit,
    vectors/vectorscopes, 270
sidelock generation of vectors,
    285-293
    applications and uses,
        290-293
signal block diagrams, 4
signal generators,
    calibration-testing with
    spectrum analyzers, 164-166
signal-to-noise ratio, 137-145
    amplitude modulation (AM)
        effects, 138
    carrier-to-noise
        measurement, 138-140
    frequency modulation (FM)
        effects, 138
    noise component, 138
    spectrum analyzers, 137-145
signature analyzers, 13,
    245-264
    accuracy, HP 5006A, 260-261
    ASCII codes, 250-251
    asynchronous noise, 262
    binary numbers, 247-248
    binary-coded decimal (BCD)
        numbers, 248
    composite signatures, 263
    conversions, numerical, 264
    data transfer processes,
        254-255
    decimal radix numbers,
        246-247
    digital circuit testers, 256-261
    EBCDIC codes, 250, 252-253
    ground loops, 262
    Hewlett-Packard HP 5006A
        signature analyzers, 256-261
    hexadecimal numbers,
        249-250
    logical numbers, 245-246
    manual troubleshooting,
        263-264
    modulation/demodulation,
        modems, 255-256
    noise, 262-263

octal numbers, 248-249
    signature characters and
        operation, HP 5006A,
        257-260
    synchronous noise, 262
    synchronous vs.
        asynchronous data transfer,
        255
sine waves, 28-30
    amplitude measurement, 62
    apparent power, 29
    average value, 28, 63
    current, 28
    differentiation, 33-35
    "dirty " sine wave: faults,
        noise, glitches, 93-94
    effective voltage, 62
    fundamentals, 35
    harmonics, 28
    impedance, 30
    instantaneous values, 62
    integration, 33-35
    oscilloscope analysis, 85
    peak, 28
    peak-to-peak, 28
    phase, 28
    power, 28-30
    power factor, 29
    true power, 29
    voltage, 28
skin effect loss, transmission
    lines, 200
software, 231, 239-243
    debugging with logic
        analyzers, 13
    LabVIEW, 239-241
    LabWindows, 241-243
spatial equalization (SEq),
    multichannel sound (MTS)
    systems, 157
spectrum analyzers, 2, 3, 5,
    13-15, 126-171
    accuracy, 14-15
    ADAC converters, 15
    Advantest R4131D spectrum
        analyzers, 131-132
    amplitude measurement, 19
    Anritsu MS610B/J spectrum
        analyzers, 132-133
    applications and uses, 126,
        137-162
    baseband analysis, 145
    BK-Precision PL 5610

spectrum analyzers (*cont.*)
  spectrum analyzers, 130-131
  calibration, noise, 146
  compression point, 159
  cost vs. performance and
    options, 126, 129-137
  digital filtering, 18
  display dynamic range,
    158-159
  display systems, 127-128
  distortion, 18
  DSOs vs., 14-15
  dynamic range
    specifications, 158-159
  Fast Fourier Transform (FFT)
    type, 16, 127, 129, 137
  frequency measurement, 19
  frequency response, 150-154
  frequency span, 137
  Hameg 500-MHz spectrum
    analyzers, 129-130
  harmonic distortion, 18
  Hewlett-Packard HP 3563A
    spectrum analyzers, 137
  Hewlett-Packard HP 8590B
    spectrum analyzers, 135-136
  Hewlett-Packard 83620A
    spectrum analyzers, 168-170
  IFR Model A-7550 spectrum
    analyzers, 132-133
  input protection, 128
  intermodulation distortion,
    159
  LED readouts, 128
  manually-dialed center
    frequency types, 127
  max scan mode, 19, 137
  microprocessor-controlled
    types, 127
  modulation, 159-162
  needs analysis: selection
    criteria, 127
  noise, 18
  noise measurements, 18, 128,
    145-150
  overdriving spectrum
    analyzers, noise
    measurement, 149
  parallel-filter, 16
  performance parameters,
    14-15
  phase-lock types, 127
  portability considerations, 127

  price vs. performance, 14-15
  quasi-peak detectors, 128-129
  satellite transmissions,
    GSTAR 4 sample problems,
    142-145
  scanning per division mode,
    19
  sensitivity/range of spectrum
    analyzers, 19, 128
  signal generators,
    calibration-testing, 164-166
  signal-to-noise measurement,
    137-145
  storage capacity, 128
  superheterodyne, 17-18
  sweep generator testing,
    166-168
  synthesized type, 127, 129
  Tektronix 2710 spectrum
    analyzers, 132, 134
  Tektronix 2712 spectrum
    analyzers, 170-171
  Tektronix 492PGM spectrum
    analyzers, 136
  Tektronix 492PGM spectrum
    analyzers, 162-168
  third-order intercept (TOI),18
  time measurement, 19
  total harmonic distortion
    (THD), 146-148
  tracking generators, 128
  tuned filter, 16-17
  types of spectrum analyzers,
    127
  waveform analyzers vs.,
    20-21
  YIG oscillators, 15
  zero scan mode, 19, 137
square waves, 30-32
  applications and uses, 32
  duty cycle, 30-31, 64
  fault testing, 32
  frequency measurement, 64
  generation of square waves,
    32
  harmonics, 31
  oscilloscope analysis, 55,
    63-64, 85
  power, 64
  risetimes, 31
  symmetrical display, 31
standards, communications
  standards, 232-234

state functions, logic analyzers,
  12
storage devices/media, 231-232
stored writing speed,
  digital-storage oscilloscopes,
  98
superheterodyne analyzers,
  17-18
surface-wave acoustical filters
  (SWIFS), 42
sweep generators, spectrum
  analyzers testing, 166-168
sync signals, 45-47
synchronous data transfer, 255
synchronous noise, 262
system return loss, 214

T

Tektronix, 3
Tektronix 1503C reflectometer,
  208-217
  8-bit microprocessor, block
    diagram, 210
  applications and uses,
    209-215
  calibration, 208
  capacitor added to line, 214
  controls, 211
  dielectric-contingent velocity,
    215
  display systems, 212-213
  features and performance,
    208
  impedance matching, 214
  impedance measurement,
    212
  length measurement, 212
  millirho measurements,
    215-217
  operation, 208-209
  printouts, 210-211
  reflected voltage, 213
  setup, 209-212
  system return loss, 214
Tektronix 2232 digital-storage
  oscilloscopes (DSO),
  104-113
  Acquisition mode use,
    111-112
  analog portion of the 2232,
    106
  channels, 106
  dual storage display, 111-112

frequency-difference measurement, 110
front panel/controls described, 105
memory capacity, 109
NTSC test pattern, 107-108
one-shot capture, 106
photography, camera attachment, 106
price range vs. performance quality, 104
sampling rates, 108-109
staircase signals, 112-113
Storage command use, 109-110
storage modes, 108-113
time base accuracy, 106
trigger selection, 106
video multiburst waveform, 107
Tektronix 2252 programmable oscilloscope, 70-75, 88-95
block diagram/component description, 88
counter-timer, hardcopy, display, 75-77
deflection factors, 72
description of the 2252 and options, 71
development, 70-71
front panel showing switches/functions, 73
mode, cursor, time-base controls, 75-76
modes, mode buttons, 72
photograph or print results, 72
sweep parameters, 72
trigger, slope, channel, coupling selection, 74
triggering controls/functions, 72
Tektronix 2710 spectrum analyzers, 132, 134
Tektronix 2712 spectrum analyzers, 170-171
Tektronix 492PGM spectrum analyzers, 136, 162-168
accuracy/precision, 162
dc input blocking, 164
Help button, 163
impedance matching, 164
pulse analysis, 163-164

signal generators, calibration-testing, 164-166
sweep generator testing, 166-168
triggering, 163
Tektronix TSG170G vectors/vectorscopes, 267-269
telephone faults, fiberoptics, 227
terminal fault locators, optical reflectometers, 228
termination characters, communications standards, 233
test patterns, oscilloscope, 83
test signals
  transmission lines, 200-201
  video signals, 47-50
Think Jet printer, 116
third-order intercept (TOI), spectrum analyzers, 18
time base requirements, 1, 7-8, 65-70
  digital-storage oscilloscopes, 121
  HM 604 time base oscilloscope, Hameg Corp., 66-70
time constants
  long time constant, 35
  LR circuits, 32
  RC circuits, 32
  short time constant, 34
time-domain reflectometry (*see also* fiberoptics; optical reflectometers; transmission lines), 194-228
applications and uses, 209-215
basic procedures, 194
capacitance line terminations, 202-203
capacitor added to line, 214
characteristic impedance, transmission lines, 200
circuit-board connections and connectors, 205-208
complex impedance, transmission lines, 199
complex-load calculations, 198-200
controls, Tektronix 1503C, 211

dielectric-contingent velocity, Tektronix 1503C, 215
distance to discontinuity, 195
impedance matching, Tektronix 1503C, 214
impedance measurement, Tektronix 1503C, 212
input impedance, 196-198
large reactances, 201-204
length measurement of cable, Tektronix 1503C, 212
limitations, 204
long line TDRs, 208-217
lossless (ideal) transmission lines, impedance, 199
millirho measurement, Tektronix 1503C, 215-217
optical reflectometers, 217-228
padders, 204-205
printouts, 210-211
probe connections, 205-206
pulse measurement, 194
reflected voltage, Tektronix 1503C, 213
resistance loads vs. waveshape, steps, 195
resistive discontinuities, 201
risetimes, 204
setup, Tektronix 1503C, 209-212
skin effect loss, 200
small reactances, 204
system return loss, 214
Tektronix 1503C reflectometer, 208-217
test signals, 200-201
voltage reflection coefficient, 196, 201
timing functions, logic analyzers, 12
total harmonic distortion (THD), 146-148
modulation, 160
tracer fault locators, optical reflectometers, 228
training/operator competance, 2-3
transformers
  coupling circuits, 39-40
  efficiency, 40
  impedance, 39-40
  maximum transformer efficiency, 40

transmission lines (*see also* fiberoptics; time-domain reflectometry)
capacitance line terminations, 202-203
capacitor added to line, Tektronix 1503C, 214
characteristic impedance, 200
complex impedance, 199
complex-load calculations, 198-200
dielectric-contingent velocity, Tektronix 1503C, 215
distance-to-discontinuity, 195
ground loops, 262
impedance matching, 214
impedance measurement, Tektronix 1503C, 212
input impedance, 196-198
large reactances, 201-204
length measurement, Tektronix 1503C, 212
lossless (ideal) line, impedance, 199
millirho measurement, Tektronix 1503C, 215-217
padders, 204-205
reflected voltage, Tektronix 1503C, 213
resistance loads vs. waveshape, steps, 195
resistive discontinuities, 201
skin effect loss, 200
small reactances, 204
system return loss, 214
test signals, 200-201
time-domain reflectometry, 194-228
voltage reflection coefficient, 196, 201
triggered sweep oscilloscopes, 1
triggers
differentiated voltages as triggers, 35
glitch trigger, digital-storage oscilloscopes, 102
logic analyzers, 11-12
troubleshooting, basic troubleshooting steps, 4-6

true power, sine wave, 29
tuned filter analyzers, 16-17

U

useful bandwidth (USB), DSOs, 10, 99

V

vector analysis
vectors/vectorscopes, 265-294
applications and uses, 266-267
complex vs. real numbers, 274
gated rainbow vectors, 84-86
j operators, 272-273
Lissajous patterns, 274-280
NTSC-type, 265-269
origination of vectors, 269-274
oscilloscope, 84-86
out-of-phase sine waves, different amplitudes, 273
phase alternate line (PAL) system, 269
phase change measurements, 265
phase shift, calculations, 280-282
phase shift, illustration, 282-285
polar equations, 271
rainbow generators, 84-86
series-tuned circuit, 270
sidelock generation, 285-290
sidelock generators, applications and uses, 290-293
Tektronix TSG170G vectors/vectorscopes, 267-269
troubleshooting uses, 267
vertical blanking interval (VBI), 81-83
vertical interval reference signal (VIRS), 48-53
vertical interval reference signal (VIRS), 81-83
vertical interval test signal (VITS), 50-53, 81-83

vertical resolution, digital-storage oscilloscopes, 100
video signals, 43-53
black reference, 45-47
blanking, 45-47
color burst, 44
IRE references, 45
NTSC signals, 44-47
sync, 45-47
test signals, 47-50
vertical interval reference signal (VIRS), 48-53
vertical interval test signal (VITS), 50-53
white reference, peak white, 45
VMEbus, 234
voltage, sine wave, 28
voltage reflection coefficient, time-domain reflectometry, 196, 201
voltmeters, 1
VXIbus, 233-234

W

waveform analyzers, 2, 19-23
applications and uses, 19-20
dynamic range, 21
frequency range, 21
HDTV technology, 21-23
Hewlett-Packard HP3562A Dynamic Signal Analyzer, 20-21
memory requirements, 21
resolution bandwidth, 21
spectrum analyzers vs., 20-21
waveforms, 27-53
analysis of waveforms: basic procedures, 27-28
coupling circuits, 37-40
filters, 40-43
pulse circuits, 36-37
sine waves, 28-30
square waves, 30-32
video signals, 43-53
waveshaping, 32-35
differentiation, 33-35
hyperbolic functions, 32-33

integration, 33-35
time constants, 32
wirebond connections,
    fiberoptics, 227

X

X/Y board, oscilloscope, 59-60

Y

Y controls, oscilloscope, 58-59
YIG oscillators, 15

Z

Zenith ADC 2300 U
    multichannel sound (MTS)
    system, 155-157
Zenith APU 2400 U
    multichannel sound (MTS)
    system, 155-157
Zenith BTSC-dBx multichannel
    sound system, 154-158
zero scan mode, spectrum
    analyzers, 19

Other Bestsellers of Related Interest

POWER SUPPLIES: Switching Regulators, Inverters & Converters—*Irving M. Gottlieb*

This book is a comprehensive guide to operation power sources used in applications from computers and radio transmitters to TVs, and more! It contains the details and depth required by electronics professionals and the basic explanations and advice needed by hobbyists. This book offers a wide range of related technical data in a single format. 448 pages, 260 illus. **Book No. 1665, $21.95 paperback only**

BASIC ELECTRONICS COURSE—2nd Edition
—*Norman H. Crowhurst*

Absolutely no previous project building experience is necessary to assemble these devices! Projects include a pendulum clock, a siren, a music box, a photocell-activated night light, an audible continuity checker, a proximity detector, a one-IC AM radio and an electronic noise maker. This volume will introduce you to TTL and CMOS ICs, FETs, SCRs, Triacs, IR transmitters and receivers, and many other devices beyond the ordinary bipolar transistor. 400 pages, 347 illustrations. **Book No. 2613, $17.95 paperback, $24.95 hardcover**

HOW TO USE SPECIAL-PURPOSE ICs
—*Delton T. Horn*

A truly excellent overview of the newest and most useful special purpose ICs available today, this sourcebook covers practical uses for circuits ranging from voltage regulators to CPUs . . . from telephone ICs to multiplexers and demultiplexers . . . from video ICs to stereo synthesizers . . . and more! Easy-to-follow explanations are supported by drawings, diagrams, and schematics. 400 pages, 392 illustrations. **Book No. 2625, $16.95 paperback, $23.95 hardcover**

SOLID-STATE ELECTRONICS THEORY WITH EXPERIMENTS—*M. J. Sanfilippo*

This learn-by-doing approach to solid-state electronics makes even the most advanced concepts amazingly easy to grasp! With an emphasis on practically, it covers everything from the simple two-junction transistor to state-of-the-art GaAsFETs (Gallium Arsenide Field Effect Transistors) and opto-electronic devices . . . from basic-state theory to practical, hands-on applications. 336 pages, 325 illustrations. **Book No. 2926, $16.95 paperback, $25.95 hardcover**

LEARNING ELECTRONICS: Theory and Experiments with Computer-Aided Instruction for the Commodore 64™/128™—*R. Jesse Phagan and William Spaulding*

Exciting and unique, this book is a self-contained manual for student-managed learning. There are quizzes throughout and two main exams that will help chart your progress and pinpoint areas that need more attention. Answers are provided. 360 pages, 273 illustrations. **Book No. 2882, $16.95 paperback, $24.95 hardcover**

FIBEROPTICS AND LASER HANDBOOK —2nd Edition—*Edward L. Safford, Jr., and John A. McCann*

Explore the dramatic impact that lasers and fiberoptics have on our daily lives—PLUS, exciting ideas for your own experiments! Now, with the help of experts Safford and McCann, you'll discover the most current concepts, practices, and applications of fiberoptics, lasers, and electromagnetic radiation technology. Included are terms and definitions, discussions of the types and operations of current systems, and amazingly simple experiments you can conduct! 240 pages, 108 illustrations. **Book No. 2981, $18.95 paperback, $24.95 hardcover**

LEARNING ELECTRONICS: Theory and Experiments with Computer-Aided Instruction for the IBM —*R. Jesse Phagan and William Spaulding*

This easy-to-use guide combines electronics theory and hands-on practice with computer-aided instruction. Designed as a self-study guild to be used in conjunction with an IBM PC or compatible, this book is perfect for the electronics student or hobbyist. Computer programs are provided to graphically show concepts and mathematical relationships. 340 pages, 271 illustrations. **Book No. 3082, $16.95 paperback, $24.95 hardcover**

UNDERSTANDING LASERS—*Stan Gibilisco*

If you could have only one book that would tell you everything you need to know about lasers and their applications—this would be the book for you! Covering all types of laser applications—from fiberoptics to supermarket checkout registers—Stan Gibilisco offers a comprehensive overview of this fascinating phenomenon of light. He describes what lasers are and how they work, and examines in detail the different kinds of lasers in use today. 180 pages, 96 illustrations. **Book No. 3175, $14.95 paperback, $23.95 hardcover**